Praise for *Flavor Flimflammery*

"As a ruler who does not even tolerate naked emperors, I certainly reject flimflam on any level so therefore welcome this royal tome." —King of Floonia, psychologist

"I hated it. It made a fool of me with all those facts, science, common sense, and inventiveness. What happened to being a slave to conformity? And who is this emperor who is underdressed?" —Margorie Bloop, wine (not water) sommelier

"Gerry's book is certainly more entertaining than what I used to find in his jeans pockets when he was a boy. I dreaded having to wash them. Worms." —Minnie Nicholls, Gerry's mother

Figure 1.
"I'm picking up delicate, ambrosial notes of a briny, oleogustus, piquant, savory fusion flavor."
"So, it's a cheese sandwich then."

Christina Swanke, illustrator and cartoonist, is so much more than Gerry's wife and a steadying force. Here is how Christina describes herself.

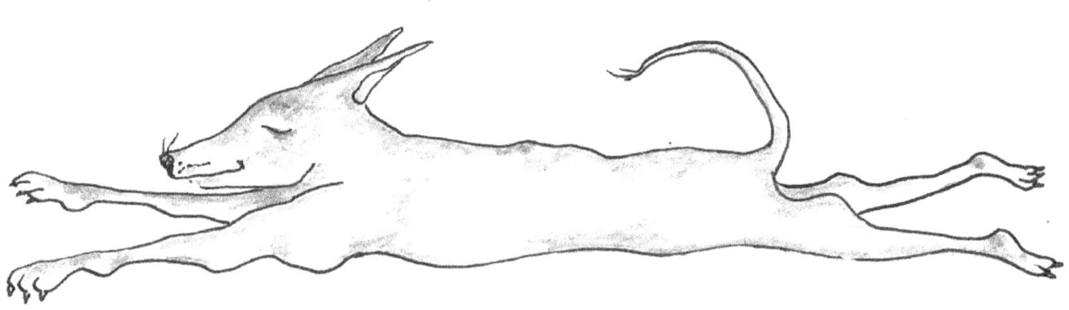

"Living with another creative soul and a lumpy dog has been a wonderful voyage. The capacity to be passionately curious about the world is a gift, I think. Hence, my joy is found somewhere between cultivating moss, painting with watercolors, making kefir and growing food; all of which require tinkering."

An epitaph for the eventually-deceased Gerry Nicholls:

According to legendary keyboard-player-to-the-stars Bob Noble, Gerry Nicholls wears many hats. Despite not having any official medical qualifications, he has an encyclopedic knowledge of all aspects of healthcare. (Many doctors regard him as their go-to second opinion.) Gerry is also quite funny. His outlandish sense of humor and spirit of reasonable contention make him a spectacularly original, nonconforming voice. This, assuredly, is a strength—the man should not be underestimated.

All love letters and strongly worded critiques can be directed to
Gerry Nicholls. Special thanks to Chole, our esteemed Director of
Artistic Restraint, for keeping things (mostly) in check.

http://www.flavorflimflammery.com

Flavor Flimflammery

By Gerry Nicholls GED, HYBC

with Dr Paul Connolly PhD Psychology

PALMETTO
PUBLISHING
Charleston, SC
www.PalmettoPublishing.com

Hardcover ISBN: 979-8-8229-3974-5
Paperback ISBN: 979-8-8229-3975-2
eBook ISBN: 979-8-8229-3976-9

Flavor Flimflammery

FLAVOR PERCEPTION IN HUMANS, TEENAGERS, AND OTHER CREATURES

By Gerry Nicholls GED, HYBC

with

Paul Connolly, PhD (psychology),

USCG Licensed Captain, First Prize, Eighth Grade Science Fair,

Volunteer CT Audubon Osprey Watch,

King of Floonia

and

Christina Swanke, illustrator

TABLE OF CONTENTS

KNOW THAT:

This book is titled *Flavor Flimflammery* for a reason, so lighten up.

"Flimflam" has negative connotations, which include fraud and deceptive nonsense (according to Webster's dictionary), so it is without apology that I name names. If I do not name names, then this would be incomplete reporting or bad science. You could accuse me of not doing a thorough job if I hid their identities to protect the fraudsters. Despite all attempts to tone it down, the result was watering down the humor—especially my Brit humo(u)r.

The downside is that I tend to come across as negative and a whining moaner.

It became apparent that a frequent part of this book involved the psychology of why people engage in negativity and resent being called out on it, so I took the easy way out and engaged Dr. Paul Connolly, psychologist/humorist, to tease out the depths of human behavior for me.

For the purposes of this book, fraud includes those supposed scientists who doggedly refuse to accept that a principle of science allows them to be wrong with the production of fresh information.

Penn & Teller are the ones to trust in the flavor world and the world in general. As internationally, massively successful magicians, they have embraced skepticism as part of their ethos. You should understand that magicians are by nature honest because they let you know in advance that they are going to fool you—unless you prefer to believe that they did make that building disappear. Penn & Teller are also science- and mathematics-based, but it is your job to dig deeper if you wish to understand why and how. Worth doing.

It is fine to fail to understand the significance of some of my theories.

You don't understand flavor perception? Don't worry; most people don't even understand pain, yet 30 percent of the world endure it chronically.

If professional flavor people accepted my definitions unconditionally, they would lose money. The more flimflam, the more money.

Flimflam—crossword-puzzle followers use *bilk, bamboozle, hoodwink, fleece, dupe, nonsense, deceive,* and *scam* as synonyms.

There was much, much more really interesting stuff I could have included, so forgive me if I didn't embrace your research.

The Role of Humor in Flavor Flimflammery

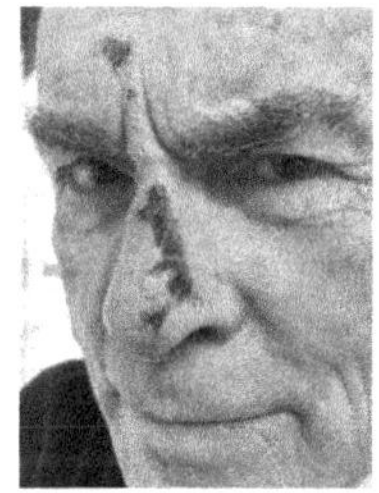

It became quickly apparent that one cannot expose flimflam without "putting some peoples' noses out of joint."

In this book, I often combine facts, science, and humor to accomplish this. All three of these can really bug certain people, so this is where I reveal that I am a Brit, and we are infamous for many traits, which include "Brit humor" as well as bad teeth. This may be described as the fine, indelicate art of mockery, which in our case, we may impolitely call "taking the Mickey," or more commonly, "taking the p**s" ("p**s-taking"). This may be interpreted by non-Brits as being unnecessarily blunt or even cruel. It is common for us Brits (and others) to make our point by taking the p**s. It is what we do. Having said that, I am happy to report that it has not only spread to Australia and South Africa but also to France and even Germany (yes, the latter was an example of taking the p**s out of the Germans). During World War II, there were even hit songs taking the p**s out of Hitler.

Regarding this book, it has been pointed out that people in general prefer humor to hate, so how should I get my point across without appearing to use my writing to execute a grudge? The answer: use Brit humor.

People who I respect.

Who else managed to finesse their way into this book by sneakily using honesty, intellectual curiosity, and stunning good looks?

Maria Geismar—raised in the Azores under third-world conditions regarding water and electricity and arrived in the USA speaking no English yet progressed to becoming a recovery room RN specialist. She volunteered on our first medical mission to Ethiopia (with Doc Rob) to support the Ethiopia's Daughters nonprofit. Maria is loaded with common sense and kindness and is a book reader. Being retired now, she offered to read the entire first draft text and be blunt with her comments. Perfect.

Dr. Rob Weiss, MD, director, and founder of CT ENT Sinus Center, Hearing & Balance, has created one of the preeminent medical practices in Fairfield County, Connecticut. (In the text he is referred to as Doc Rob.) Doc Rob came on our first medical mission to Ethiopia with just a small medical bag and a big trust that I knew what I was doing. Over the years, he observed my interest in flavor perception and how smell loss was not a subject that even ENT doctors had training to treat. With the advent of COVID-19 smell loss, he was able to utilize my knowledge and advice to educate himself and his staff. What Doc Rob is a master of is living his life to the full, physically, and mentally.

INTRODUCTION

Hello, my name is Gerry. Nice to meet you. I am known for thinking outside the box. Aren't you glad I didn't start by thanking my agent and publisher? That's enough; you can continue by jumping ahead and reading the book now.

Figure 3. Dog's best friend, the author.

What is my point?

Let's get straight to it regarding the mystery of the flavor world, such as how other flavor explanations fail to express this subject.

Smell exists in a physical form—real molecules.

Taste exists in a physical form—real molecules.

Flavor does not exist in a physical form. It is all concocted in the brain.

As you will read later in this book, you cannot put flavor into a gas chromatograph for analysis. The same implies to entering the word "delicious" for gas chromatography analysis. Yes, this therefore means that flavor and taste are not interchangeable words. Sit tight; flavor is a phantom concocted in the brain, much like humor, love, humor, jealousy, humor, irony, and humor. Okay, I admit, I like humor.

BEGINNINGS

It has been said about me, "Gerry Nicholls can be described as an annoying, sesquipedalian, polymath maverick and an iconoclastic, inveterate skeptic." Let's make it simpler. It would be easy to cast me as a complaining old crone. Easy, because it would be broadly true. I am quick to point out that it does not mean I am simply a grumpy old man with a dull axe to grind. On the contrary, I am simply a seeker of truth and someone who likes to laugh while he accomplishes this. My task is to explain flavor perception in its broadest sense using original thinking, hard science, and common sense and lace everything with buckets of humor. My show-business background shines through in every page if I have done my job. I have no fear of offending people if I sense that I am right, and they are wrong. The book explains smell, taste, and flavor perception in a whole, original, fresh way that branches off into the seldom-charted territory of ancillary subjects, which include flavor wheels, astringency, wine-tasting limits, nonlethal weapons, dogs' paws, gas chromatography, garlic breath—a real A to Z, finishing with A: anosmia.

How to Read This Book

What's my point?

There is a flow to the book. Treat it like any story.

Let's make this brief. When you read (past tense) the collected works of William Shakespeare, I bet that you started at the beginning and worked your way forward, didn't you? If you started in the middle, then there is no hope for you. Think of this book as a kind of life, eating, and drinking story. I shall be referring to concepts previously mentioned, so absolutely start on page one, and sally forth without fear, fury, or delay. No cheating—promise? And don't trust me or believe anyone. Except Darwin.

Don't Read This Section

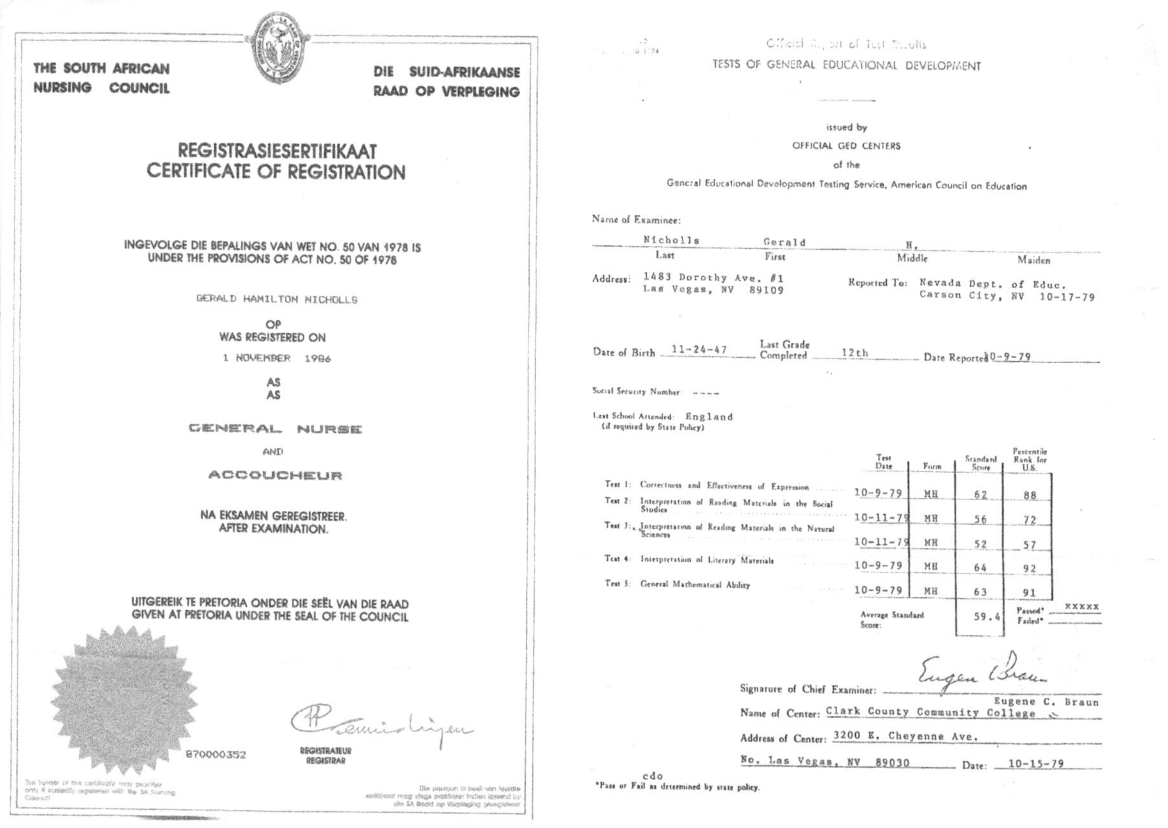

Figure 4. Real qualifications as an expression of free thinking.

I know I just told you to read everything from beginning to end, but this section is all about me. Me, me, me. Skip to the next section—I won't be offended. But you can't, can you? Because you are inquisitive and highly intelligent. If I had a PhD, then I could write glowingly about myself, but I do not have such a qualification. I have various qualifications, but the qualifications I am most proud of are my GED (not General Eating Diploma but General Education Diploma) and HYBC (Hundred Yards Backstroke Certificate), plus my *registrasiesertifikaat* confirming me as an accoucheur (see figure 4). Don't tell me you've never heard of that before, *meneer*? A male midwife, to you. The GED qualification, I earned without preparation whilst on vacation in Las Vegas from South Africa. This is where a lesser writer might name-drop. I might even have been sober, too, but I cannot swear to it. The point is that I believe having a PhD in sensory sciences has shown to be a negative attribute. I was going to include my CPR certificate until I found out that CPR essentially does not work, especially in the way and level of effectiveness that we are led to believe. This book is not about CPR though. It's about flavor perception, so using the same level of skepticism regarding the aforementioned, I have come to question all sorts of established wisdom regarding eating and drinking. What I have discovered shocks the heck out of me—hence this book. But it is not just me. In my view, the ultimate in science-based cooking is *The Food Lab* by J. Kenji López-Alt (nerd alert—it is nearly a thousand pages long).

If you are sitting comfortably, then let me tell you how things started. In 1990 (after South Africa took away my passport and I sued them and they let me leave), in the very early days of specialty coffee, I went to a roaster, where they offered a couple of coffees described as "full-bodied" and "robust." I asked the barista what the difference was. No answer. The instructions for coffee making included the advice to use fresh, cold water. You see this advice to this day. Why? Why not skanky, old, hot water? You are only going to heat it up anyway. That is how I started with what I now refer to as my "kitchen science experiments," which are the backbone of this book. You are getting tired, so I have addressed the world of Gerry's kitchen science in much greater depth later in the book. Skip ahead if you are impatient to test me. Here's a teaser: it has connections to the hit TV series *Cheers*. I knew that would get your attention, unless you were too young to have wallowed in this funny series.

Apologies and Disclaimers

What's my point?

It is not unreasonable for me to expect a lot of pushbacks from what I say in this book.

Strange how truth has the ability to raise hackles.

If you are looking for a flavor anatomy/physiology book, then this is not it. My experience is that very, very few people are interested in those aspects of flavor perception. Of course, I must address anatomical structures and physiological mechanisms, but I do so in a fresh way—poor in detail but rich in concepts. For example, when I talk about the trigeminal nerve, and I do ad nauseam, I actually do realize that the large and intimidating vagus nerve is nearby and is also collecting and sending data. The trigeminal nerve for me is a concept that I name "mouth trigeminal perception."

Homeostasis

In a nutshell, this word explains the optimal functioning of the body typified by processes such as body temperature and blood pressure—anything that keeps the body on an even, healthy, and efficient keel, allowing us to procreate and live a long and healthy life. Throughout the book, I repeatedly use *homeostasis*, because it is, after all, why we don't keep keeling over dead more often than we already do.

Accuracy

In this book, concepts are more important than getting into too many details, so expect me to concentrate on broad brushstrokes, which will apply in many directions and areas. Concepts are difficult, especially when one must devise them oneself—original thinking, in other words. I had an offer from a world-famous science writer to help me with this book (a steal at $25,000 up-front) but I had to turn him down, because he didn't know the concepts.

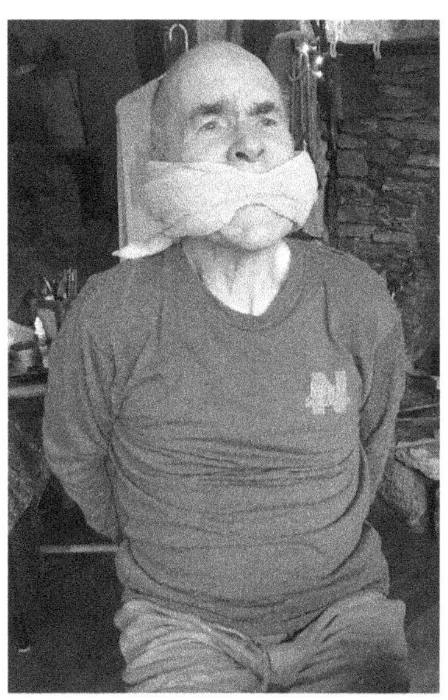

Figure 5. This is how many sensory scientists would like to see me.

Admittedly, I have not been in touch with every major name in the sensory sciences connected with flavor, but I get the picture from those with whom I have been in contact. They really don't want Gerry Nicholls (who does not even have a PhD) muscling in on their little club. I am not saying that they would like to see me tied to a chair to keep me quiet, but what is clear is that when I ask a question or propose a theory that they did not conceive, they tend to respond by running away. Hold on Gerry; they are scientists. Where is their intellectual enquiry? Where is the inquisitiveness? None. Run away and stop answering his annoying emails. No polite "Dear Gerry, I disagree..." No, they simply rudely ignore me. Sad, isn't it? They discredit the scientific process and community. They have not advanced science by their lack of genuine interest in flavor science.

Many people will not like my writings. I may embarrass and offend people. Sad to say I have found that many scientists will even respond to my questioning with what is effectively, "Leave me alone; I don't want to talk about it"—even famous-in-their-field sensory scientists. I will do them a favor by not revealing them by name. It is not my intention to offend people, but if, in the pursuit of truth, I do, then so be it. Truth always wins out. There are enough books easily available which cover technical, detailed aspects of smell, taste, and flavor. They talk about nerve pathways and regions of the brain, so I won't need to cover that. Finally, I hope that you will have noticed that I have not put a bunch of recipes at the end to pad things out. I put in a Notes page instead. Please take that as a sign of strength and a solid book (unless it is a cookbook).

What's My Point?
My point is that many authors just don't get to the point in the sensory sciences.

Some just randomly go from one piece of research to the next and lazily rehash information. They don't engage in original thinking and concepts. But in steps Gerry Nicholls. Apart from looking forward to a massive influx of royalties over the coming years, the book is designed to entertain you, make you laugh, and most importantly, make you think about and question everything in life. Don't believe anyone. Starting with a downer, I must confess I have been highly disappointed by the lack of intellectual curiosity by flavor scientists. Having been on taste panels for beer, cider, honey, wine, and coffee for years, I have had to restrain myself when I hear flowery descriptions of the drinks. The trouble is that I tried standing up, puffing out my hairy chest, harrumphing, and declaring, "You are talking hogwash." All it got me was black looks and no respect—not even when I justified my views scientifically and logically. Even though I admit to being an overopinionated know-it-all, if you can explain to me why I am wrong by using logic and science, then I will happily back down. This happened once when I gave a talk in the UK that included the concept of "supertasters."

The head brewer, Adrian Redgrove of UK's brewery Castle Rock, told me that "supertaster" sounded elitist. I removed it from future talks. He was right, and I was wrong. Since then, I have learned that Adrian is generally right about these things. Damn that British common sense.

Do We Really Need Another Taste Book?

Unfortunately, yes! Sure, there are plenty of books out there on smell, taste, and flavor in humans (and dogs), but none of them attempts to tell the full story—to give an explanation of what smell, taste, and flavor are. Personally, I believe that is an impossible task, but I will try to give you a striking feel for the possible world of flavor perception. If I am successful, you will say, "I didn't know that there was so much to it." The recently published book on flavor handily titled *Flavor* is by science writer Bob Holmes, who goes into great depth regarding hot peppers but does not attempt to get to grips with what smell, taste, and flavor are really about for humans and why we are different from other mammals.

Neuroenlogy by Professor Gordon Shepherd is a wonderful book on the science of wine tasting, which goes into great detail on retronasal smell, for instance. Retronasal smell is a concept which people rarely care about or really understand. Even the author agrees with me on this as indicated in personal communication. There is no comprehensive handbook on smell, taste, and flavor perception, nor a tome where one can consult metallic taste, flavor wheels, and anosmia.

The massive industries of wine, beer, and coffee tasting all follow a similar, too-well-trodden path on sensory analysis. Prepared to make enemies, I will state something that will upset people: the sensory analysis of drinks needs a complete overhaul. There, I said it. You may sit down now. Pour yourself a drink. If one accepts the premise that we all live in our own flavor world, then it is a short journey to say that trying to be specific or definitive regarding flavor perception is a fool's errand.

In my experience, the massive wine, beer, and coffee worlds just want to keep the status quo, in part because they don't see a problem existing. If they admit to it, then they are in for a long and very expensive rewrite of everything they have been espousing and selling for decades. Should they decide to revamp their sensory analysis courses, then the ramifications will be huge. Let's take one small example. Most use flavor wheels. Incidentally, flavor wheels are used by all sorts of specialties, including brewer's malt, tomatoes, honey, cheese, and, hilariously, body odor. The World Coffee Research organization recently spent $200,000 on developing their flavor wheel and sensory lexicon.

Being a presenter for the Specialty Coffee Association of America, I am expected to use their information. Is there a problem, Gerry? Yes, I have many problems with their information. Take one simple example. They insist that lime is bitter, and lemon is not. They seem to have confused bitterness with astringency. If they get that wrong, then there's no hope for the rest of what they are saying (and selling and selling and selling). Later in this book, I address flavor wheels in general and in detail, and I can't wait; you are in for a treat. It's so funny to me, but not to them.

Why Didn't Someone Else Write This Book?

What's my point?

They are not as skeptical, brave, and freethinking as me.

Flavor perception's hard to understand.
It's not just some spices and carbohydrates, bland.
The way it's explained just drives me mental.
Gonna say my piece, and I won't be gentle.

There is so much nonsense, like flavor-wheel foolery.
I have to say, it's useless—it's really no tool to me.
Food-pairing flimflam—what's that all about?
Duck liver pâté with red wine will only give you gout.

With science, skepticism, and rational thinking,
you'd wonder why I leave the experts blinking.
They all should shut up and read this book.
and revise their flavor-perception outlook (nod to Dr. Seuss).

Where's the science, honesty, common sense, and intellectual curiosity? You would think that because, as of last count, humans mostly eat and drink every few hours, we would know just about everything there is to know on the subject. Eating and drinking have consumed our thoughts and drive since we split from other apes. Why didn't some PhD sensory scientists write this book? Why me? I am retired and don't need money, fame, or glory. Full disclosure, I quite like glory.

Writing a book for the intelligently curious
should make some of them extremely furious.
Science is science; a fact is a fact.
Seeking the truth 'til the subject is cracked.

I have come to believe that sensory scientists are often stuck in old habits. They are not prepared to question everything. They are not Penn & Teller aficionados prepared to challenge their bosses, organizations, and education and are highly unlikely to be members of the Skeptics Society. Someone said, "Be wary of someone who has nothing to lose." Therefore, don't trust me—I may tell the truths which you don't want to hear. Finding potential flaws in perceived truths regarding sensory perception takes no great effort.

Let's take a specific example. Bob Holmes, the aforementioned science writer who wrote the wonderful book *Flavor*, writes about an oft-repeated concept which has appeared in most sensory analysis books. As he says, "Some say smell accounts for 70 percent of flavor; others put it at 90 percent or more. A source calling itself Livescience says 80 percent of taste is smell." That is even wrong, because they mean flavor, not taste (basic stuff).

Flavorjournal.biomedical. says smell accounts for 75 to 90 percent of flavor. University of Florida Center for Smell and Taste says 80 percent of flavor is smell, although in their defense, the director told me face-to-face that it was nonsense. To have a percentage, one must have a unit of measurement. There is no unit of measurement for smell, taste, or flavor. So, at best, one can say, "Some of flavor is attributable to smell." Beyond that, nothing. Hold on; Bob Holmes is a science writer for *New Scientist* journal—the biggest-selling science journal in the world. I contacted Bob Holmes on this, and he (reluctantly) agreed with me. You see, there is so much perceived wisdom—just perceived. Oh yes, that means that a GED man can teach a PhD person basic science/math principle.

My bête noire is the oft-repeated statement by people who talk about metallic taste and smell. Seeing that almost nothing has been written on it, I can be forgiven for thinking that would be a quick task. It turns out that after asking just about everyone I meet, most have experienced "metallic taste." All well and good, except that metallic taste does not, and cannot, exist except as a linguistic brain phantom invented in your brain, and it cannot be transferred to mine. Okay, change the laws of physics and chemistry if you must. In this book, "metallic taste" is a longish chapter (ten pages at last count) on just this subject in chapter 6 ahead.

The implications are that I am left fearful (and curious) as to what I might discover when I dig skeptically and deeply into the rest of the flavor world. So, my point is that in the sensory science world, more education does not equate to an increased ability to explain a subject to the public, even though I was told by a famous authority that I had no standing because I was not a known academic. This is where I want to scream that Galileo did not even have a basic degree, let alone a PhD. It's Mr. Galileo to you. Hold my coat, please, as I roll up my sleeves. It saddens me that so many sensory scientists are not genuinely, intellectually curious enough to broaden their horizons and be open to the truth. This attitude hinders the evolution and advancement of flavor perception. For 99 percent or more of us, others' higher education in this subject does not amount to a hill of beans. And there's another thing: What's a hill of beans? And here's yet another thing: more education will not make any of them funny, although eating too many beans may help.

Gerry's Kitchen Science Experiments

What's my point?

You don't need tightly controlled conditions to do useful science.

As mentioned previously, the backbone of this book is my kitchen science experiments. It all started with coffee and the instructions to make coffee with fresh, cold water. The Penn & Teller in me asked, "Why?" There were no books to consult, and there still aren't any twenty-five years later. Let me just name-drop here by mentioning Charles Darwin and how he did garden science in much the understated way that I do my kitchen science. He also was obsessed by how emotions affect us, so we have lots in common, although he was admittedly a lot smarter than me. As a national beer judge, I applied some of our blind-tasting techniques to the problem. Using numbered wine glasses and a standard weight in grams of coffee grounds in each glass, I set forth to simply alter the temperature of the water poured in them. Here's what I did:

- Water at boiling point
- Water at 205°F
- Water at 195°F
- Water at 180°F
- Water at 170° F
- Water desaturated of oxygen.
- Water super-saturated with oxygen (yes, I used a 100 percent oxygen infuser)
- Water—regular faucet drinking water
- Water demineralized (distilled)
- Water at 205°F, straight out of the cold tap and heated up
- Water at 205°F, old, skanky, and straight out of the hot tap (or faucet, for us Americans)

Inviting the beer judges was easy. Those were the lovely old days when foreign, interesting beers were hard to acquire. Most of the tasters were also coffee drinkers. So, what were the results? It was all too easy for words. Broadly speaking, as long as the water was very hot (above 170°F) and rich with minerals, then nothing else had much, if any, effect. The big—no, BIG—lesson was how much information I got from a simple test in my kitchen. I was sure that doing the same at a university sensory sciences lab would not reveal much extra information, if any at all. Don't forget, all I wanted to know was if it was necessary to use cold, fresh water. I detected flimflam. The reason I altered the oxygen concentrations was because oxygen dissolves easier in cold water. You knew that. I remembered that from physics class at Colchester Royal Grammar School in England. Colchester is England's oldest town, and we had to wear the traditional school uniform, which consisted of a glowing, royal-purple, peaked cap with matching blazer, gray shorts with matching gray socks, and black shoes. This is all longhand for, "Bye-bye girls; run away!" No one had a girlfriend, except Gordon, who had a yacht. Not hard to work out. However, I digress. Let me travel at warp speed forward to 2017 in England again.

Castle Rock Brewery in Nottingham, England, engaged me to consult with them on producing a coffee stout. No, you can't just splash some supermarket hot coffee into beer. They knew I had previously helped Connecticut's largest brewery to produce a (actually, not *a* but *the most*) fabulous coffee stout. They also knew that Two Roads Brewing had a full-time chemistry lab on-site, and despite this, I insisted on doing things my way, just like Frank Sinatra. This included educating the executive team, including the CEO, what coffee is before it even gets into the beer. I insisted that their laboratory use my kitchen science techniques. Was I successful? Just go to Russia, Scandinavia, Germany, England, and good old North America to sample Espressway Stout and find out.

Enough about how wonderful I am. The point is that we take lots of eating, drinking, smell, taste, and flavor knowledge for granted. Oh, so you want an example. How about this: Do food and drink folk know lemon from lime? That's right; can you tell lemon from lime because you use them both every day? Wrong. No, you probably can't. How do I know? Gerry's kitchen science. How is your journey to hate me coming along? One thing I wondered is how kitchen science stacks up against professional sensory science lab science. In other words, would I be better off using a proper lab even though my gut said no? If you have never seen a sensory science lab, here's what one looks like:

Figure 6. Question: Was this photo taken at a sensory science lab or a post office or a prison visiting area?

Just like eating at home, isn't it?

They will tell you that they have control over as many variables as possible in order to be objective. I would say that's great, except that these are not the conditions of how people consume food and drink in the real world. This is exactly how people never consume food and drink. People consume with every distraction imaginable. Enter a bar like the one in *"Cheers"* (the real one being in Boston), but here in Connecticut, we have the next best thing: MY PLACE located in Newtown, Connecticut. It has many of the attributes of the hit TV series *"Cheers"*, except for the international notoriety.

Figure 7. This is a real-world sensory lab.

The regulars at the bar are my non-ad-hoc taste panel. It is a joy for me to call out to Sharky the barman, "Hey, Sharky, what's the difference between lemon and lime?"

Figure 8. Sharky squeezing lemon or lime.

"I don't know, so leave me alone."

At the end of the book will be revealed some of the eye-opening and shocking results of my kitchen science, some of which were performed at MY PLACE (see figure 7).

Sensory Scientists: Why They Fail in the Real World

Let me be blunt. (You mean, you haven't been up till now, Gerry?) Ultimately, all flavor perception is about science in some form. The same goes for much of life. We are bombarded by endless "studies," but little attention is paid to the world of the scientists who produce them. Scientists are just people with all the foibles that we all exhibit. Previously, my broad image of all scientists (including my brother) was gentle, passive, heads-down-in-the-laboratory, talking a language that only they fully understand. It was only when I started writing this book that I discovered a darker side. Scientists by and large deal in minutiae of facts and observations. Scientists need to specialize. That means they learn more and more about a narrower and narrower sub-subject until they know everything about almost nothing. My current view is that flavor perception scientists not only tend to lose the big picture but don't even know or care that a big picture exists. Exceptions occur, of course. I hereby state that flavor sensory scientists are the wrong people to explain flavor perception to the masses.

Let me repeat that louder:

Flavor sensory scientists are the wrong people to explain flavor perception to the masses.

Let's face it; they have failed to explain flavor as it applies to every living human being. That should rattle a few people. My next awakening was that professional flavor perception science is beholden in large part to the massive food and drink industry. Realize that all the foregoing and the information and views hereafter are sweeping generalizations. Exceptions occur everywhere, but let's call them exceptions, not the rule. Here's a lighthearted way of putting it: if you are a stand-up comedian, just hope that there are no sensory scientists in the audience, because if you say something like, "A man walks into a bar," they will want to know what the bar was made of, its temperature, its height above sea level, and so forth. Don't explain that you are joking, because they might not comprehend that either.

Figure 9. Is this your image of the average scientist?

I wouldn't blame you if you think that this is extreme and unfair. However, I wouldn't agree with you though, from my experience. It disappoints me greatly that I even have to write this chapter. Let's get something straight: I have no "official" science qualifications but have followed, studied, and been enthralled by all sciences on my own since I left school. Nerd alert! There was a time in England I would arise at 5:00 a.m. to watch *The Open University* on the BBC. I couldn't get enough and still can't. Science, for many of us—for all its faults—is all we have. Science, of course, as with everything in life, has its exceptions. For me, David Attenborough, Steven Hawking, and Neil DeGrasse Tyson are the first to come to mind. Just so I don't maximize my enemy list, I consider that for this book (only) there are two science disciplines: research science and analytical science. Research science is typified by medical drug studies of population groups, and analytical science is what Carl Linnaeus did with taxonomy and the classification of organisms. I classify sensory science as research (especially sensory science research) but let us first investigate what drives this research.

Research Science: What's It All About If Not Science?

A pal of mine is a research chemist, and I am left with the belief that he thinks that flavor perception can be explained solely by chemicals and receptors. Yes, I know that everything in the universe(s) is made up of chemicals. Despite this being true, if he is correct, this still would not explain how someone might detect a smell when there is nothing there. Synesthetes do this all the time. It seriously unsettles me when I must reveal the following: most research scientists are under pressure, not in the pursuit of truth, but in the pursuit of funding.

Funding, funding, funding, funding, funding, funding, funding.

Money, money, money, money, money, money, money.

Pandering to funders, pandering to funders, pandering to funders.

Keep my job, keep my job, keep my job, keep my job, keep my job.

Another friend of mine (yes, I have more than one) is an analytical scientist at a famous museum. His wife is a research scientist, and I invited them out for drinks. I even said I would pay. She declined, because in her free time, she was too busy writing grant proposals. That's a highfalutin way of saying "begging." He admitted being jealous of me because I have no need of funding, I have no boss, I have no schedule, I have no deadlines, and I can write whatever I want, plus I am under no pressure to rush out lots of research papers (forget about long-term studies). Yes, I have plenty of money too. This goes a long way to explaining the attitude, mood, and tone of this book. Sometimes I call it, "My manual of truth, unencumbered by financial and destructive influences." Please don't use the word *bribery*. Medical drug researchers especially hate to hear that word, despite it being demonstrably true of late. Let's face it; the end result should be truth rather than a smile on corporate sponsors' faces. At this stage, I expect you to be horribly depressed and disappointed like me, but never fear—future chapters I bet will have you laughing your heads off, or I have failed. In a strange quirk of reasoning, a scientist did a study and showed that most research is eventually shown to be wrong. Of course, his research could be wrong too. Let's stop there. It will come as no surprise that sensory scientists as a group/club/clique/galère would be very happy if I did not exist. Is there something about me that repulses them? No.

Galileo, I think it is safe to say, is universally considered to be quite a smarty-pants. In his time, he came up against much opposition to his thoughts and theories. I think they liked his beard though. Darwin receives flack even today. Let's time travel and zoom ahead to today. Last week, the *New York Times* magazine ran a feature on bird feathers. Richard Prum, PhD, of Yale Peabody Museum was heavily featured with his bird- and dinosaur-feather revelations. Rick and the Pulitzer Prize go hand in hand. He likes feathers, and I challenge you to find anyone who understands their evolutionary implications better than him. He is world-renowned and respected and was recently given a huge bag of money by some foundation to say, "Thank you, Rick, for your contributions to science. It's all yours. Buy a boat if you want, or just fritter it away." He in fact did and sailed into the sunset. Yeah, right. Despite the foregoing, even Rick receives pushback from some of his peers who are reluctant to help him. They don't reply to emails or phone calls. As the *New York Times* article states, "Prum still stings from the perceived scorn from his academic peers." Of course, I would tag on words like *jealousy* and *resentment* in there, but that's just me.

Let's travel back to a month ago, to the 22/29 December 2018 edition of *New Scientist* journal (starting on page seventy if you just happen to have it in front of you on your lap). The article was a cheery Christmas article about scientists arguing. It also contained fascinating information on the electrocution of an elephant. Those British. Scientists can't resist turf wars apparently. Astrophysicists don't like mathematicians tramping on their physics turf. Geologists don't appreciate paleontologists getting in on their act. "Stick to your rocks!" Sensory scientists don't like Gerry Nicholls having any sort of opinion and upsetting their sensory science club—especially if he uses science to back up his ideas. When I told a distinguished professor at a university a couple of states west of Kentucky (I'm doing my best to be cryptic and not to embarrass him, even though he deserves worse) that the flavor wheel and sensory lexicon he was a consultant on and which had a budget of $200,000 was claptrap, he was unhappy. Actually, *claptrap* is not strong enough a descriptor. Mind you, I was very sneaky, because I used a device to expose him called science. To get a deeper sense of the situation, forge ahead and read my chapter 7 on flavor wheels and metallic taste. Referring to the *New Scientist* article, a Nobel Prize–winning physicist was "rappelled" down a peg by a non-professional scientist yet a skilled mountaineer, who had real-life experiences with the geological formations to which the physicist had referred. The mountaineer said, "All you have is a sexy story. I have run my raw, frost-bitten fingers on those rock formations." Touché.

WHAT IS SMELL?

What's my point?

Everyone knows, but it is hard to explain.

Remind me again the difference between aroma, nose, olfactant, bouquet, fragrance, scent, odor, and *smell*? If you are describing food and drinks, do not complicate things; just use the word "smell," and we will all understand and get along a lot better. Addressing a group of doctors to guide them on how they should advise their patients who have smell loss, I opened with a question: "What is smell?" Not one doctor answered.

When one is involved with official tastings of wine, beer, coffee, cheese, honey, olive oil, and endlessly on, the assessment forms innocently (or otherwise) try to make it as complicated as possible. I think that the major problem is that very, very few people understand what smell is and does. Likewise with taste and flavor. If you are judging a wine, beer, or coffee, you will probably be instructed to sniff the sample as bouquet, not aroma. It is much scarier than that. Consider wine. If you want to appear a real wine smarty-pants, you should hold your mammoth glass by the base and sniff. No chance of spilling, because the wine only occupies a small space at the bottom of the glass. No laughing or mockery, please. Give off an air of aloofness and all-knowingness. Temper this with holier-than-thou notes. Sniff—oh, sorry to use such a base word, nose—and let your imagination act as though it is on LSD. Disclaimer: that's exactly how I do it (without the actual LSD), but at least I know I am engaging in flimflam.

Why do we smell the wine? Because it is fun. No other reason. So, what's your point Gerry? The main reason to smell anything is to decide up front if you want to continue with things and proceed with swallowing. This concept was covered in detail in chapter 3 ("Gerry's Three-Gate Theory").

Bringing you back to earth, on a practical note, if you are a coffee importer and have a shipping container of green coffee on offer, it's a good idea to know that the shipment that you ordered lives up to your expectations. In practice, small samples are taken randomly from 60 kg bags, roasted lightly, and ground. The importer smells the grounds. If she smells a defect—don't get me started on potato-taste defect, which devastated the coffee economy in Rwanda—then she can reject the shipment then and there with no need to bother making a sample from it. It does not make her universally popular, but she doesn't care. I once cupped a coffee sample which smelled of blue cheese. Reject. Smelling is fun but not for everyone.

Ahh, the wonderful smell of a celebratory dinner. I mean, for 95 percent of us. The other one in twenty humans has some degree of smell impairment or disorder. If you are keen to freak yourself out, jump ahead to the section on anosmia. Don't. Leave the depressing news until the end of this otherwise cheerful book.

My Old Granny Smells

Lots of creatures smell (transitive and intransitive verb). By and large, they do it to locate food, mates, or danger. That includes us, although whereas we use our noses, flies use their legs, and large-brained dolphins just could not give a care. Dogs? I just read *Being a Dog* by Alexandra Horowitz. It is okay to read it to your dog. On an evolutionary level, we don't smell in order to experience pizza. Pay attention; why do we smell (transitive verb)? From an evolutionary standpoint, we smell to ensure us a secure place in our environment, not to maximize our wine enjoyment. That may have shocked you. How do we know this is true? Ask people who have lost their sense of smell. The overriding symptom is depression. Boo-hoo? I don't think so. Without getting morbid, the singer from the band INX hanged himself. Ask COVID-19 patients who have suffered anosmia and lost their sense of smell. It is now considered an international health crisis. Assuming that you like your home, when passing through your front door after work, the home smell map in your brain tells you that you are secure and loved (we hope). You don't think about it consciously. You are home, so relax and feel secure. What's for supper?

"I smell gas!"—not secure. "I smell the kitchen stove on fire!"—not secure. "I smell supper!"—secure.

Smell: Two Roads to the Brain

Smell is not one thing, one process. It is two—sort of, most of the time. Here, we come to my bête noir (actually I have several, to be honest with you). Traditional sources of smell explanation rely on the words *orthonasal* and *retronasal*. Almost all standard literature uses these terms. My temples are throbbing now as my blood pressure creeps up to levels only Tour de France cyclists can be proud of. To be clear, from here on, smell is derived from actual, real, physical molecules, which I will call *olfactants*. So, one smells thanks to olfactants.

Orthonasal Smell

What does "ortho" mean? Some version of straight, upright, right, or correct. Wow, that is so helpful, isn't it? Now we can consider orthonasal smell, or straight nose smell, or upright nose smell, or right nose smell, or finally, correct nose smell. Is it just me, or is this flimflam nonsense, especially as we have a single word that requires no explanation to replace orthonasal smell? You'd better stay seated because this next part is an intellectual mind blast. Wait for it—why don't we just use the word...wait for it...*sniff*? Just to cover our bases, the opposite of *sniff* is *snort*. We sniff in and snort out (I know, except with cocaine). Dogs snort a lot; if you don't have a dog, rescue one.

Retronasal Smell: Just Oft-Repeated Ignorance?

What does "retro" mean? Broadly, "from the recent past or backward." Retronasal smell is also sometimes called "mouth smell." The word *retronasal* makes no sense to me, and up goes my blood pressure again. Professor Gordon Shepherd (to his credit) in his book *Neuroenology* used "internal smell" as well as "retronasal." This still does not give a clear picture of the process. I spoke to Professor Shepherd on the subject, and he agreed with me. He then asked me how I explain it. I told him, to which he responded, laughing, "Gerry, this is all very good and fun." Luckily, and I hope deservedly, I have a friend, Doc Rob (ENT). His specialty even includes the word "nose". He was eager to elucidate.

"*Retronasal* doesn't mean anything to me either. In daily medical practice, we have no need to use this concept, but knowing Gerry, I thought we could discuss it over a small glass of wine to stimulate creative thinking. Well, anyway, that was our excuse. We discussed what the process was in practical terms that could be compared to something else in life which everyone would know about. Let me reveal our thought process:

You swallow the item, which is rich in olfactant (smell) molecules. By swallowing, the item is forced backward down the throat and into the stomach, but not before a plume of olfactant molecules wafts up high into the nasal cavity. Yes, to the same exact location as the sniffed molecules went. One destination, two routes. We then challenged ourselves to answer: In life, what liquid process goes crashing down and sends a plume of itself upward? Answer: A crashing wave. To make it easy for you, here's an illustration:

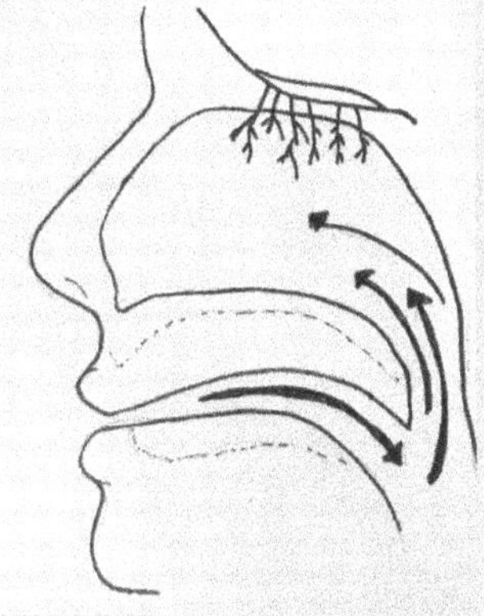

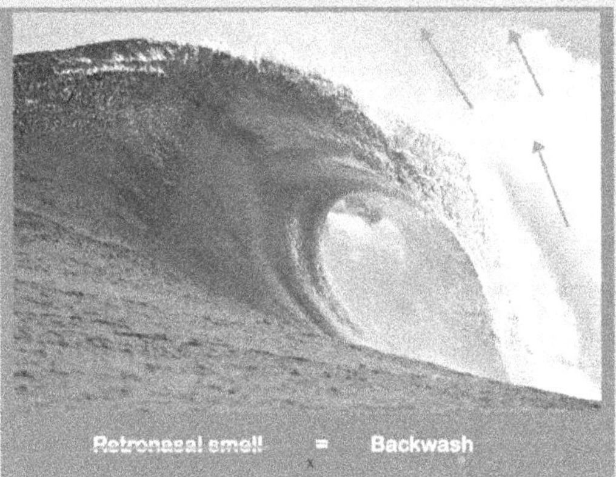

Figures 23 and 24

In Hawaii, surfers might refer to this as "backwash"—but there again, they might not, because surfers have their own cool language. Regardless, "backwash" seems a vast improvement on the other word, which we won't repeat. Having tried out "backwash" on audiences, I can tell you that after leaving the talk, it has become part of their vocabulary".

Medical Students: This Is for You

Understanding the brain does not have to be complicated if you stop thinking of it as it is usually depicted—a brain. If you read standard works on flavor perception (and I encourage you to if only to emphasize why buying this book was such a brilliant, inspired idea), tucked in the text is the inevitable journey into the human brain. How the world loves to read about the prefrontal cortex and associated structures. Perhaps not. Writers cannot resist talking about the olfactory epithelium, cribriform plate, olfactory bulb, and so on, inward, and upward. An ear, nose, and throat doctor friend said to me in a locker room, "We don't know anything about the cribriform plate and stuff like that." Exactly. Why would you expect a nose doctor to know anything about noses?

Why Sensory Professionals Fail in Explaining Smell

At the SmellTaste 2017 conference at the University of Florida, I studied the audience during a forty-five-minute flavor perception presentation given by one of America's most respected sensory scientists. She lost the audience. They didn't understand, and they didn't care that they didn't understand. I do understand that I used "didn't" three times in one sentence. After I asked awkward questions from the floor, a doctor sidled up to me by the fruit bowl—in a manner more reminiscent of how spies approach each other—and he asked secretively, "Who the heck are you?"

"Trouble," I answered. If I'd have been living in Gainesville, Florida, he would have offered me a job, he proffered.

Back at home in Connecticut, tucked in my home in our oak forest, I decided to try a whole new tack to explain flavor perception—throw out all the (failed) old supposed science but retain the core science principles. It dawned on me that I have worked in the operating room with many neurosurgeons. As surgeons, they need to be precise, we would think when they talk about neural anatomy. If so, do they talk about a hirsute covering of the epidermis, which overlies the subdermal lipid layer, which itself envelops the osseous cage encapsulating gray nervous tissue? No, they use the word *head*, or if pressed, *brain*. Given this, why do sensory scientists wax lyrical about cribriform plates? If neurosurgeons can get to the point with just *head* and *brain*, why can't I come up with something that isolates the smell, taste, and flavor perception parts of the brain? Here is the region we are considering (figure 25):

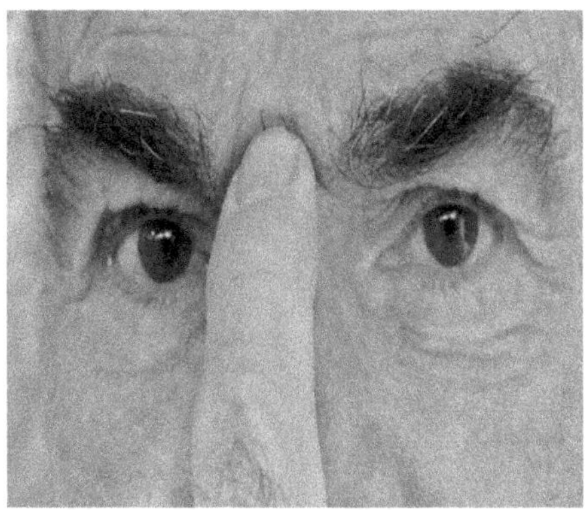

Figure 25

At the top of the nose, roughly between your eyes, is where all the smell molecules go, right about here—but on the inside of course.

Gerry's Original, Appropriately Named Anatomical Region

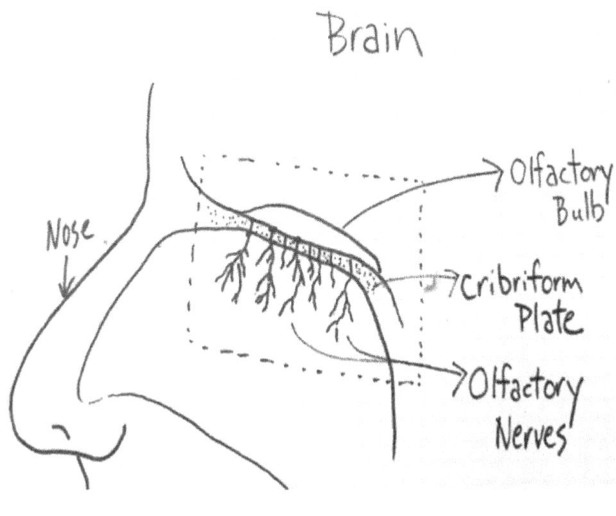

Figure 26

If you could poke your finger up your nose, stopping just short of puncturing your brain, you will be tickling your olfactory epithelium. It's not a big piece of tissue—the size of your fingernail, depending on how big your fingers are. But you get the point. (By the way, descriptions like this certainly grab the attention of the average fifteen-year-old.) The olfactory epithelium is a little piece of delicate, octopus-like tissue which hangs out of the brain, ready to snag its next treat of olfactant molecules, like fingers hanging down. When I speak to medical professionals, I get them to mimic this (see figure 27):

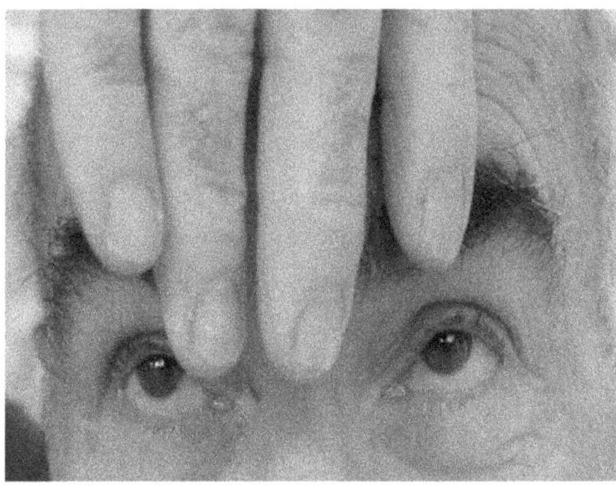

Figure 27

Are they offended? No chance, because they have so few opportunities to be silly and have fun, unlike the nurses and technicians.

In a desperate effort to break the tension of you trying to grasp this, here's an aside which you might find insightful and entertaining. Back in the locker room, a surgeon was complaining about his life, which involved his wife spending too much money, enabled by him being on call too often. It was easy to get him to see reality: "You're pretty smart, doctor; why don't you become a nurse and get relaxed and happy?" Here's the great part: he took me seriously and just replied, "I don't think that my wife would go for that." I'll let you read between the lines there.

Now, where were we before you interrupted me? Yes, having fun. Look at figure 27 again, which shows the olfactory epithelium in finger form.

Figure 28. Olfactory epithelium to brain: "It's really lovely outside today."

All your guts/intestines have access to the outside world, too, but only that one tiny part of your brain shares that access to the outside world. The guts are basically just a donut hole through your body from gob to anus. Sorry about the language, Angel. I didn't mean to say "gob." In the USA, the word *piehole* is used, so I believe.

The three structures (cribriform plate, olfactory bulb, and olfactory nerves) to me are screaming, "This is smell central, where it all starts to happen. Together, we make up a visual structure, and all we need is a name to define us. Make it something easy, like *brain* or *head*, for goodness' sake." As I tucked myself in for a restful sleep that night, I gave myself a final demand: make the name understandable by any fifteen-year-old science student. Scrap that—make it understandable by any fifteen-year-old art student. So, what shall I call this grouping of anatomical structures? (See figure 29.)

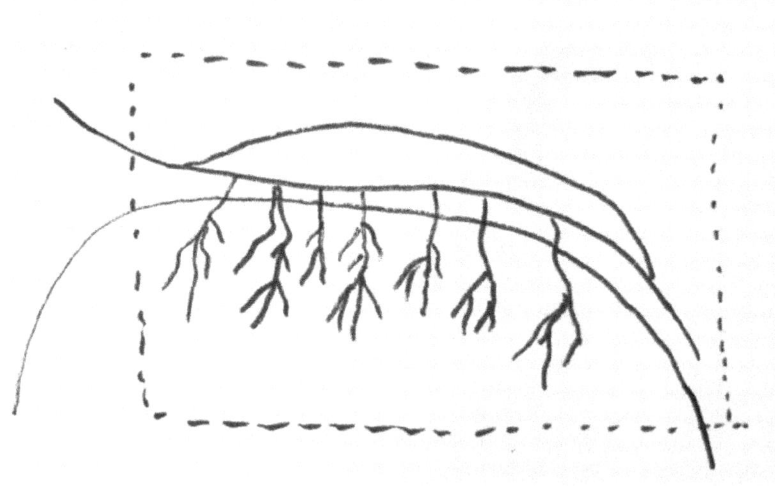

Figure 29

My brain would not let my adenosine neurotransmitters (to activate sleep) kick in until it agreed to one more demand: pretend that no one had ever tried to describe smell, taste, and flavor perception before. You have guessed it; this was a recipe for waking at 3:00 a.m. with the cartoon light bulb flashing over my head and the sound of me screaming out, "Eureka!" At 2:30 a.m., off went that light bulb. One problem: I did not want to get out of bed into the cold of my library to put pen to paper. This scenario has played out before, whereby at 3:00 a.m., the idea is solid but has evaporated by 7:00 a.m. I got up. Surprisingly, as dawn light arrived, the idea was still established in my brain. I had it all worked out.

Dear reader, you have no idea what is coming, but believe me, it is good. Really good. Instead of *head* or *brain*, my explanation for all flavor perception was a word just as short. My epiphany was the "bit of brain hanging out" part. At 3:00 a.m., I expressed it thus: when you get down to it, the olfactory epithelium is just a bit of brain hanging out. Obvious, isn't it? Bit Of Brain Hanging Out. *BobHo.*

This arrangement reminded me of something that we experience in everyday life. Tiny packages of olfactant molecules go up to a region which is really a neurological sorting and processing department. BobHo analyzes the information and decides where in the higher brain regions to send it. Isn't it a bit like the postal service? Packages go to the post office sorting area, where decisions are made on where to redirect the packages to their destination.

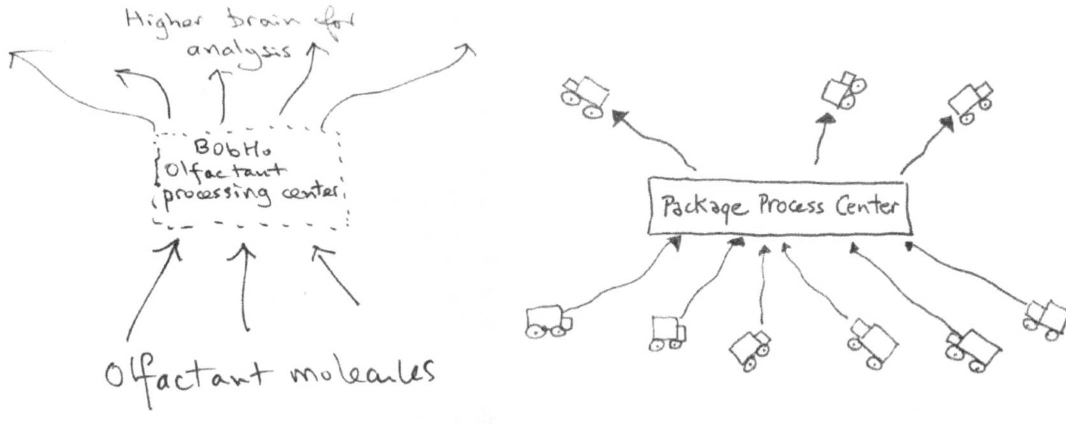

Figures 30 and 31

Let's humanize that even more: BobHo. BobHo has the answer to it all. Good old Bob. Then, to take it a step further...

Dogs, Bees, and Wives

Let's lighten things up with answers to questions I often receive. What about dogs, bees, and our wives when it comes to smell acuity? How good is our sense of smell compared to dogs and bees and wives? When it comes to smell acuity, you cannot compare wives to bees to dogs. Bees are not interested in ovulating dogs, dogs don't need to seek out flowers, and although wives may seek out flowers, it is not to eat parts of them. Jim's wife is self-protective by smelling his marijuana so she can stop any of his misguided plans to move into the shed for a quick hit. It never takes long when discussing how good we are as smellers before the subject of perfumers and sommeliers arises. It is romantic to fantasize about a perfumer's abilities. They are born with this amazing talent, yes? No. Are they gifted? The definition of "gifted" is universally not agreed upon. Young students being accepted into a gifted program can be tested using standard IQ tests. There is no test for smell, taste, and flavor, and there isn't a perfumer or sommelier whose first words were not "Mama" but "notes of oak moss" or "dark fruits." It is all learned. Bob, my keyboard-player-to-the-stars friend, was labelled as gifted at age three. Bob thinks that "gifted" means he simply found playing piano easier than others and he enjoyed it. No magic gift. Is being handsome a gift? (Yes, but for other reasons.) Supposedly, gifted people have one thing in common. They have a natural interest/passion in a subject, but to express it fully, it always requires lots and lots and lots and lots of hard work. It's the same for top Olympic skiers or chess players. Should you wish to scare yourself, watch one of the documentaries on sommeliers training for their exam. Not many people are inspired to arise at 4:30 a.m. to drink wine before work. How about top chess players? One could reasonably say that the world champion is gifted, but at what stage are lesser players not gifted but simply very good? Where is the line?

To smell anything, molecules of olfactants must enter the nose (or mouth). How many molecules does one need to identify the smell? I have no idea and neither does anyone else. What is known, though, is that some smells require relatively few compared to others. We describe this as us being "sensitive" to a smell—how many molecules it takes before we can detect and name it. Let us think out of the box and consider some of our other senses: sight, hearing, and touch. Do we ever say, "She is such a great seer," or, "He's an amazing hearer"? Now you are going to bring up the subject of people with perfect pitch. They surely have some basic degree of innate natural ability, but after that, it is just work—just like the perfumers and sommeliers. Let me say it: the first utterance of a child has never been "B-flat." Sure, some of us are naturally better at certain things in life than others—a certain giftedness. One thing is for sure: many of you will disagree with me on this. My wife pointed out to me that in some cultures, they consider intellectual disability a gift. People with Down syndrome are revered.

What about touch? It's no good; I am not going there. You just work it out for yourself.

CHAPTER 3

GERRY'S THREE-GATE THEORY

What's my point?

To satisfy your urge to consume a food or drink item, one must go through some decision processes.

You don't usually just grab whatever is closest to you. Let's start by considering dogs. You put a bowl of food down for Bonzo, and what is the reaction? If he's not hungry, he will reward you with indifference. If he's hungry, he doesn't just chow down without a pause; he needs an introductory sniff. If the food passes the sniff test, then all pretenses of discretion are cast aside, and it is consumed without fanfare or much chewing or thanks. With humans, things are somewhat different. It is my pleasure to give talks to the food and beverage industry in all its broadness. Restaurateurs, wine sommeliers, bar servers, mixologists, and waiting staff all hang on to my every word of wisdom. "Yeah, right," I hear you say again. It struck me that customers don't just wander in from the street to some bar or restaurant unthinkingly and order the first thing on the menu. Humans go though some sort of vetting, selection, and decision process before putting gastronomic consumables down their throats—one throat per person. This progression of decision-making led me to my internationally long-awaited "Three-Gate Theory of Consumption."

Gate 1 allows one to progress to Gate 2, which is a commitment to at least try the offering by smelling it. Gate 1 says, "So far, so good; I like the look of this locale/offering, and all things being equal, I will consume it at Gate 3 after confirmation by Gate 2."

Gate 1

Figure 10. "What are you in the mood for? Crowds and loud music or awkward silence between us?"

Selecting any bar/restaurant at random, the first decision is whether to enter the locality at all. Yes, or no?

If one seeks a quiet drink and a refined meal, then the local sports bar is probably not going to entice you through Gate 1. Bright lights, loud sound systems, and TV monitors are not what you are after. Naturally, one can turn that around and say that the thirsty sports fan is unlikely to satisfy her needs at Café Chez Montseuriont de la Fille, Paris. Just peeking in the window may lure you inside but hold your *chevaux*; there are more decisions to be made before you order the deep-fried witchetty grubs or French country goat cheese platter.

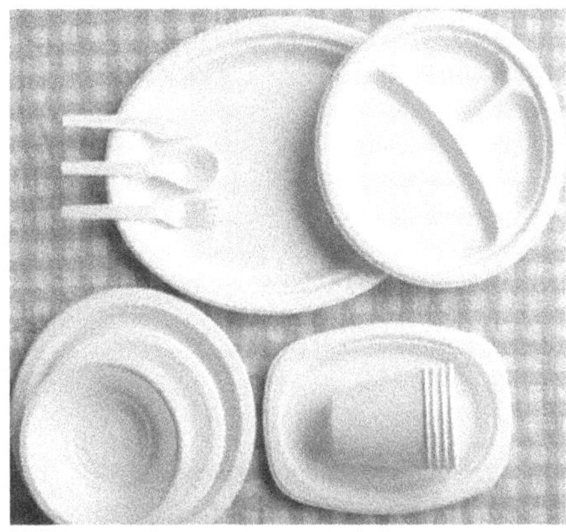

In the marketing world, there are endless variations on the old concept of, "Presentation, looks, first impressions, and appearances are everything." Well, here's the thing: they are darned right. Why? It all comes back to my precept that all flavor perception revolves around emotion. We'll get to that in exciting detail later. Relax, for now. The stronger the positive emotion, the better, but it does not take much to jog that to negativity—for me, bright lights, sports on TV screens, and an unhealthy menu, and tag on to that a dirty toilet. None of what I am saying is earth-shatteringly complex. Consider also that the attitude of the staff, the interior décor, and naturally, the menu offerings are strongly Gate 1. Gate 1 may stop you in very subtle ways. Once, I gave a smell-loss (anosmia) lecture to doctors in a hospital, and afterward, Doc Rob, ENT, offered to buy me a breakfast of eggs. I accepted. He bought them. I declined. "Why not?" he said. "Gate 1: the plate and cutlery," said I.

Figure 11. Gate 1 closed.

He understood because I had just been lecturing on it. For me, plastic cutlery and Styrofoam plates were unacceptable.

Gate 1 comes in many forms, including:

All Labels

Perhaps the most obvious examples are those on wine and beer bottles. Ask any liquor store owner, and it will be confirmed that customers drink with their eyes. What they wouldn't give to reduce their selection by 90 percent. There is a beer in Connecticut which is called the Naughty Nurse Amber Ale. Over the past eighteen years, it has become a huge seller. Talking to the brewer, I inquired whether he thought that if he had named it Septic Tank Ale, it would have become such a raging success, even though the beer would be identical. You can guess the answer. For me, the funny thing is that the Naughty Nurse was the nickname of a male, ultra-long-distance runner who was a midwife (accoucheur) in Africa of all things. Shh…don't tell anyone.

Names

Names can make or break a product's marketing. Have you ever eaten Patagonian toothfish or Chinese gooseberry? If you think that they are not names that would attract buyers, you are right, and that is why marketing folk changed their names to Chilean seabass and kiwifruit.

Logos

Logos are powerful. Just think of the golden arches (figure 12) or the Guinness harp (figure 13). No explanations are necessary.

Figures 12 and 13

Latte Art

It brings out a warm glow to most coffee drinkers, even though it is simply milk, yet it is so much more—in fact, enough to produce international latte-art competitions.

Your Favorite Cup

Figure 14

Sets the stage for a pleasant drinking experience. See figure 14.

Door Entrance Staff (Don't Say Bouncers)

Figure 15

This is Phil. You cannot help but like Phil as he asks for your ID at Two Roads Brewing. I am seventy-six years old and don't mind at all. He considers his job to be to entice people in, not keep them out. Performing his job, he makes useful recommendations regarding the beers and places to sit. He is polite and has a huge welcoming smile. Presentation is everything. He knows this and is proud that he is Gate 1.

Bad Apples

Figure 16

Supermarket food displays are Gate 1. Bad apples physically, visually, and emotionally taint good apples. Food stores expend enormous effort on presenting their goods in the best light. They take the outer leaves off cabbages and make sure there are no blemished fruits among the unblemished ones. This drives Joni Mitchell crazy. "Give me spots on my apples but leave me the birds and the bees. Please!"

Toilets

Figure 17

How obvious can it be that clean bathrooms are appreciated by customers? Yet that so frequently does not translate into reality. A store owner I know invests (yes, he thinks of it as an investment) twenty dollars a week for fresh bathroom flowers. Believe it or not, men enjoy flowers probably as much as women. Don't ask me how I know. Perhaps you are a man and are with a lady partner, and on entering, she says, "Excuse me for a moment; I have to use the restroom to make myself yet more attractive." On returning, she says, "Let's get the heck out of here."

~~Gastrophysics~~psychology

Gastrophysics, a book by Charles Spence, is a bestseller. Charles is head of Crossmodal Research laboratory at Oxford University in England. It is perfectly understandable if you have no idea what the fancy, intimidating word "crossmodal" means. It is an extremely technical subject and may be above your intellect to comprehend. With regard to the foregoing, it can be expressed as, for example, "how light and sound affect the dining experience." I told you it was technical—or perhaps simply a great example of flimflam. If you are a physicist, don't even think about getting excited, because I could find close to nothing for you in his book. Physics is broadly the science of all matter and energy. On the other hand, if you are a psychologist, then this is for you, cover to cover. Did Spence spell *psychology* wrong? Curiosity got the better of me, so I wrote to him. You can guess how thrilled he was to hear from me. My email also questioned him regarding his writing "the metallic sensation we get when we taste blood." My advice? Stay away from this man. Having explained to him that no one has ever tasted metal, he responded, as expected, with a big, fat, nonscientific silence. Mind you, he might have been busy preparing to go on one of his restaurant forays where a meal will set you back the equivalent of a pair of gold wedding rings. So how did my suggestion to him to rename his book *Gastropsychology* go down? You guessed it. I would support its title being *Gastropomposity*. That would be accurate.

In order to get the opinion of a "real" physicist, who better than a genuine astrophysicist? The sky's the limit with him.

"Skimming through *Gastrophysics*, I didn't notice any of what you would consider 'physics' in the classical sense. It's mostly biology. Technically, biology is applied chemistry, and chemistry is applied physics. Ultimately, everything comes down to physics when you really get down to the basics.

Figure 18

It is all Gate 1. In my experience, Gate 1 is a concept very readily accepted by the food and beverage industries. There is nothing mind-boggling about the theory. At BevCon, a conference where I was a presenter, I was watching a vodka representative carefully setting up his stand. He turned and looked at me and said, "Gate 1," having attended my talk beforehand. Man did that feel good. Instead of a boss giving his staff directions such as, "Clean the toilets, wipe down the bar, rewrite the chalkboard," he could just say, "Bar, toilets, chalkboard...Gate 1." Gate 1 is the opener, the preview, the taster, the menu. Think of it as encouragement to continue to...

Gate 2: Smell It

Smell is tied to decision-making.

So, you decide that you are going to proceed with consuming something, whether by entering a food or drinks establishment or even at home by raiding the fridge. Even at this stage, you are not fully committed. Gate 2 is where you inspect, sniff, and smell the potential offering. With us humans, it is nowadays not so important for survival, because foods are not commonly likely to cause us illness or discomfort as in the old days when we found a dead fish in the stream. With dogs and other creatures, it is another story.

Gate 3: Let's Eat

At last, you get to swallow it—or perhaps not. Hold on to your napkin; you may get to put it in your mouth, but that does not guarantee customer satisfaction. More decisions need to be made. First, the easy one: taste, mostly on the tongue. The tongue will give your brain information regarding sweet, salt, sour, and bitter (and umami, if you believe in that sort of thing, which I don't). The tongue requires molecular particles to touch the saliva and tongue nubbins (papillae) for this. Smell molecules won't hack it. This is where the concepts of mouthfeel and palate come into play, or as I lump them, into "trigeminal perception." Let's list a few: hot, cold, slippery, slimy, gooey, fatty, creamy, crispy, crunchy, burning, irritating, numbing, painful, tingling, prickly, astringent…this all happens in the buccal cavity (mouth) and gives you another chance to back out from swallowing—think hot peppers or fish bones. It amazes me still that people will eat very hot peppers but balk at pouring hot water in their mouth. It's the same thing, after all.

So, what is the brain contemplating meanwhile? It is piecing together information from what you have seen, heard, remembered, smelled, and expected—the brain flavor map. The microbial flora in your mouth may be working to alter, for better or worse, the potential flavor. Certainly, this happens with wine. Shall we get into a discussion on slow food? When eating slowly, one might reasonably think that one is consuming smaller mouthfuls ("less-than-fulls") and chewing and chewing and chewing. This is good. Didn't your mother tell you to drink your food and eat your drink? Mine neither. With more chewing, more surface area is exposed, and there is more tongue pushing, more body heat applied, more smells released, and more time to slip easily down the throat, and simultaneously, smell molecules are released by backwash up into the nasal regions, thus causing fuller perception and enjoyment of the item. Of course, if you are consuming something like an unpleasant medication, do the opposite, which is sometimes called gulping it down.

What happens if we swallow and we have immediate, or even delayed, regret? We need to get that substance out of our body pronto. Let's consider toxins. Our first line of defense from these is at Gate 1, when we might see that we won't like them—avoidance at Gate 1. Toxins are often extremely bitter, so at Gate 3, the tongue may alert us to spit it out. Once in the gut, nausea might ensue, but we have our final option, which many people these days call throwing up. The desire to vomit can be instigated by strong emotions (not just sick with love), medications (cancer therapies especially), and previously mentioned revolting smells, which include rotting fish. Regardless, with strong muscular contractions, we can force-eject gut contents from the stomach and further below (oh, that is what colors it yellow then). For reasons I do not understand, evolution has decided that horses, rats, rabbits, giraffes, and cows don't need to vomit. Have you ever tried to make a penguin vomit? I have, and it won't. Full disclosure, I was trying to collect stomach samples. Your next question should be, "How does it feed its chicks then?" The answer is regurgitation, which is not vomiting. Regurgitation is a gentle relaxation of certain sphincter muscles so that passively, food comes up. Think of it as reverse swallowing. Vomiting is self-preservation; regurgitating is nurturing.

Figure 19

How did I get the stomach samples? I wish you hadn't asked me. I might have left the word "gun" out of there. There is a first and last time for everything. Finally, a conundrum. Why do dogs eat grass then vomit? If you know, let me know. Don't keep it a secret.

Broadly speaking, your dog (if you don't have one, you should adopt one) sniffs its food bowl and either walks away or commits to consumption. It takes no time at all to decide, and once committed, it is full speed ahead to Gate 3: filling the mouth. With humans, it can be much more complicated and protracted. Smelling the item can give us much information about it and perhaps drag up strong emotions from our memory banks. In the case of cheese, one yellowish lump looks much like any other, but one sniff can quickly separate the French from the Irish (the cheese, and the people). Gate 1 is where you accept the idea of a cheese platter, but Gate 2 is where you may decide not to try every item on offer. If you are expecting a smelly French cheese and there is not much smell, you may reject the offer because of lack of smell. Many prepared foods smell smoky—notably, meats, cheeses, beers, and distilled spirits, like whiskey. I kid you not when I tell you that apparently someone has a business selling peat-smoked water.

Yes, water. What was ever wrong with just having a glass of water and smoking a cigarette? For me, the faintest hint of smoky closes Gate 2 firmly shut. As a national beer judge, I am sometimes obliged to rate smoked beers, and I hate them all, yet my judging skills may allow me to give a smoked beer a high score. I like the smell of cigar smoke. Work that one out for yourself, please. I can't. There must be a good memory tucked in there somewhere.

Gate 2 does not only apply to food and drink. Of late, the marketing industry has realized that retail (nonfood) stores can increase sales by releasing smells into the retail space. The process has a name: Retail Scent Marketing. This does not mean that some gentle, flowery, perfumy smell will necessarily attract customers. Athletic shoe stores even use scent marketing, although I will leave it to you to decide if they utilize the sweaty feet smells that bacterial *Pediococci* produce, and which are used to ferment some Belgian beers. Yes, certain smells are known to increase shoe sales, but the same smells may—and probably will not—attract customers to ladies' underwear outlets.

Full disclosure: I may well be writing the following to entertain myself, so I will try to keep it short.

Figure 20. *"What is the best way to attract a cheetah?"*

Sit down. While we are thinking obtusely about smells, perhaps this is where I can ask you the question in figure 20, which I bet you have never been asked before. The answer is, of course, to contract smallpox. Cheetahs are attracted to patients who smell of smallpox, because they are weak and would be easier for cheetahs to convert to lunch. Before you think that I may have gone completely off my rocker, let me reveal that I got the information from a 1906 article in the *Lancet* medical journal. In the days when doctors had limited or no access to laboratories, they had to rely on simpler and more available tools to them, like their senses of smell. Diabetes was commonly detected by the patients' sweet-smelling breath. You probably would rather I don't go into detail regarding dysentery and foot rot. Did you enjoy that little aside from food smells?

Back to food and drink. These days, for us, Gate 2 tends to revolve around our potential negative perceptions of the offering. For some, all smelly cheeses shut Gate 2. It is clear to me that somewhere in my extreme youth, I was forced by my mother to eat cucumber against my wishes. For me now, its smell is the most disgusting thing I can imagine. I even detect it in watermelons. Did I mention that all flavor perception has an emotional aspect? Icelanders have a traditional food which is essentially fermented (rotten) shark meat. It is called *hákarl*. A British professional brewer friend of mine got hold of a sealed can of it. Hoping to impress his friends, he opened it to share with them. They all ran out of the room gagging, leaving him to demonstrate what a manly man he could prove himself to be. He took a bite, then threw up. The Icelanders must have a very strong, positive emotional tie to this *hákarl* fish dish.

So, these are examples of smells eliciting strong emotions and memories. Is this why there is no universally liked or disliked food or drink? Surprisingly, not everyone likes chocolate. Hard to imagine, eh? Some foods and drinks have seasonal connections, like the spices at Thanksgiving dinner or the rich (not sweet, remember) smell of Christmas pudding. On the less-attractive side, the thought of eating asparagus is inseparable from what happens minutes later for a percentage of the population, determined by genes. Of course, I refer to that universally funny and famous asparagus pee smell: asparagusic acid. Not everyone produces this sulfurous compound, hence the genetic reference. That said, there is nothing new about it. The ever-lovable Wikipedia, to which I donate financially, tantalizes us with some potentially riveting reading:

> Benjamin Franklin described the odor as "disagreeable" in Fart Proudly, his essay on flatulence[11] whilst Marcel Proust claimed that asparagus "transforms my chamber-pot into a flask of perfume."

"To each his own," I hear you say.

Figure 21. No coincidence that I held the asparagus how I did.

In my talks, I like to show a slide of me holding a bunch of asparagus, then saying, "And now, on to the smell of asparagus in urine," whilst I sneakily check out the audience's faces. This splits the audience into two camps—those laughing and those with a nonplussed, empty stare. Great fun with husbands giggling while explaining to wives and so forth (one wife per husband). A common Gate 2 negative, certainly for me, is the smell of whisky. In times past (I hope), parents often thought that hard liquor was a cure for all sorts of maladies, never mind that the patient was only three years old. That is where the relationship between feeling ill and whisky was forged.

Figure 22. *Yup, that's me again. It is not a hard concept to understand that spoiled milk, "gone off" if you are British, can spoil a nice cup of tea or coffee. Are you one of those who have the habit of going to the fridge and sniffing the milk container before committing?*

CHAPTER 4

WHY ARE THE MEANINGS OF TASTE AND FLAVOR SO HARD TO GRASP?

What's my point?

They are not; they are just generally presented in an illogical fashion.

I knew there was a problem when doctors attended my lecture on smell loss at grand rounds education at a hospital. Believe it or not, I had to start with anatomy and physiology. Over to Doc Rob (ENT):

"Yes, Gerry, I remember. Great, entertaining talk, but you lost us with retronasal smell. Here is the gist of what is commonly proffered as an explanation: each taste quality is stimulated by specific chemicals, which are recognized by receptors on cells located in taste buds within the mouth. Taste stimuli need to dissolve in saliva before they can be detected by the receptors. Dissolved chemical stimuli come into contact with receptors located at the tips of taste buds. Most taste buds are located in papillae—the small, rounded bumps you see on the tongue's surface. Some taste buds are also found on the roof of the mouth, epiglottis, and throat. This is why some medicines taste bitter even while going down. Humans have around ten thousand taste buds, clustered in projections called papillae. Each papilla can contain from one to 700 taste buds, depending upon its location on the tongue. Not all taste buds are located on the tongue; some are found on the roof of the mouth and in the throat. Each taste bud contains about fifty to eighty specialized cells. At the top of each taste bud is a taste pore, a small opening where a few cells are exposed to the inside surface of the mouth. These exposed cells contain the receptors that detect taste stimuli. A receptor cell is activated when a taste stimulus connects with its receptors. Activated cells communicate with other cells within the taste bud, some of which relay messages about the presence of taste stimuli to nerves traveling to the brain."

WAKE UP! Can you blame the audience for zoning out? Let's make it easy. A smell molecule gets snagged by a receptor. The receptor says, "Let me send what I know up to the brain for analysis." That's it, then. Easy, eh? We can all go home now. It's all correct and written for the masses like you and me, but it does not exactly leave one with the warm and fuzzies. It seems that these institutions fail to ask themselves, "Did I get my point across?" That is why I feel obliged to attempt to do it for them. After I give talks, I chat with the audience members casually and individually afterward. They are frequently brutally honest, as when the head beer brewer at a British brewery told me, "Gerry, drop the whole supertaster segment." I realized immediately that he was right, and I did drop it. After speaking to a cross section of humanity, it doesn't take long to see emerging patterns. This is part of Gerry's kitchen science. Allow me to encapsulate what I strive to accomplish by offering you this statement: a part of science is to see what everyone else can see but think what no one else has ever said. In other words, I shouldn't have to buy a laboratory and set to work with my microscope and very expensive gas chromatograph to access the truth.

The Big, Hedonic Picture

What's my point?

Ultimately, it is all about selfishness.

We consume food and drink for degrees of pleasure, not to analyze every mouthful. We are probably the only mammal that does. Why do we even care about flavor? Perhaps because eating is probably the only thing that we do that engages all of our senses. Traditionally, the big picture of flavor is explained by saying that it is a combination of smell and taste. You have heard it over and over and over and over and over again. (Very) simply put, smell + taste = flavor, but that's like saying metal lumps + gasoline = cars, or skin + bones = humans in all their complexity. Flavor is king, but you can't pin him down or define him rigidly as you can with smell and taste. Flavor is the world of food and drink that we hedonically worship. All hail the flavor king! He is the compilation of everything around him. Let me put it this way—Gerry's explanation:

"Flavor perception" is all concocted in our brain, but there are no receptors dedicated to each flavor. Flavor is whatever your brain says it is. It can lie with impunity. It exists only as much as "funny" and "sad" exist. It is easily fooled and manipulated.

Sociopsychological Contributions to Flavor Perception

Despite cultural differences from around the globe, humans congregate around food to celebrate, mourn, and memorialize events in their lives. Perhaps this is why Anthony Bourdain's award-winning food and travel series *Parts Unknown* was so popular—because he discussed food and its socio-logical implications. He was filmed in the often-noisy societal setting, not a studio. He understood that food has social connections—an expression of sharing, and for some, an intrinsic attachment to spirituality. Put simply, food and flavor can be tied to something bigger than ourselves. The act of eating can bind us to each other, to an event particularly in childhood, reminding us particularly of positive celebrations. It can transcend us. I have friends who grew up in the Azores and Ethiopia. Despite their apparent disparate differences, consuming food is truly linked to memory and emotion with a sociopsychological component. Given the following views and impressions, would it not appear beyond facile to state that flavor is simply smell plus taste? In the USA, we regularly attribute a good start to the day with the smell of freshly brewed coffee.

Ethiopia

Elshaday Solomon, registered nurse, expresses this concept with her own experiences thus: "What people eat around the world is heavily influenced by their religion, rituals, customs, traditions, and social and economic status. People's lifestyles, practices, belief systems, values, family history, background, and lineage also influence the food immensely. The profile of Ethiopian food is very distinct. It marries together earthy, spicy, tart, sour, and pungent flavors. A base seasoning, used in a wide variety of savory and spiced Ethiopian dishes, is a blend of spices known as berbere. The foundation for all Ethiopian recipes lies within its diverse mix of herbs and spices. Common spices used in Ethiopian cuisine include ginger, turmeric, paprika, *korerima*, *koseret*, *besobela*, fenugreek seeds, rosemary, garlic, cumin, cloves, cinnamon, and *timiz*.

Culturally, most of the time, we used to eat together, sharing the same table and plate. When I was a little girl, I had a swollen finger on my right hand. For that matter, I started eating by my left hand. As a rule, it is forbidden to eat using the left hand. Therefore, I was told and forced to use my swollen right-side hand, which was so painful.

The handwashing manner in our society has rules. Elders are respected and are given priority to wash their hands. On one occasion, we had a family party. I was the youngest in the family and ordered to serve handwashing. Unknowingly, instead of serving the elders first, I started to serve the younger ones first. Immediately, everybody started shouting and screaming at me. It was my unpleasant day!"

Azores

Maria Geismar is a registered nurse with much international experience gained from medical missions. She has seen it all.

"We were poor, but rich in spirit. The land provided us with our "daily bread," and from dawn to dusk, we as a family sowed, planted, and harvested. We prayed for rain, and we prayed for sun. As a young child, I remember attending novenas (nine days of prayer) for good weather, and when good weather came, we had novenas again to show our gratitude.

Once a year, a cow was slaughtered to provide a feast for the entire village. The feast was in celebration of God's granted "miracle" that usually included return of health after a grave sickness or survival of an earthquake or a shipwreck. The sense of community always added extra flavor to the already fresh, delicious food festival.

This feast of the Holy Spirit, celebrated in June, was the highlight of the village's events. Months before the celebration, children went from house to house to gather wheat grain that was milled to make the bread. It was then used to make a soup with beef broth. The women planned the baking of the bread and sweets, and the men took care of fattening the cow and gathering wine donations. I remember the excitement as a kid of gathering the goods and always trying to outdo the other kids. We talked about "the soup," and we talked about the sweets.

When the day finally arrived, we dressed in our best clothes and wore the new shoes for the occasion. We were up before the cock crowed so we could be first in line to be served the soup. Sometimes, if we were lucky, we would even be served a glass of wine. When in the Azores, I still attend the feast of the Holy Spirit. The soup is still scrumptious, and the spirit of my childhood lives on.

Our mother made special soups when we were sick—sometimes with just a simple broth, which included scallions, parsley, and an egg. Then she poured the broth over toasted wheat bread. That was "comfort food"—and it still is. If I don't feel well, I crave my mother's soups. I love making soups for my family, and if anyone is sick, family or friend, soup is usually the first comfort measure that comes to mind. The soup pot is on. Food was revered, from the time of sowing to the kitchen table. It was a gift of Mother Nature and of God. It still is, and I remind my grandchildren that it is to be treated as such. No complaining, no waste!"

King Flavor: All Hail the King!

Figure 32

King Flavor runs his world using three main executive departments: smell, mouth (buccal cavity), and brain. "That's not what other books say!" I hear you say. Quite correct. This is where all the annoying details raise their heads. From here on, as with any bureaucracy, King Flavor has lots of subdepartments jostling for attention, so pay attention please. Put down that donut. Like any corporation and bureaucracy, they are dependent on all the workers—from the bottom up. Sometimes the departments work nicely together, and sometimes they don't. But the king is always right. He is never wrong. Now we need to look a little closer at smell, mouth, and brain. The brain is where King Flavor resides. He is the head of the brain as well as the brain of the head. Yes, you can now have a small glass of wine. I said, "small glass."

Figure 33

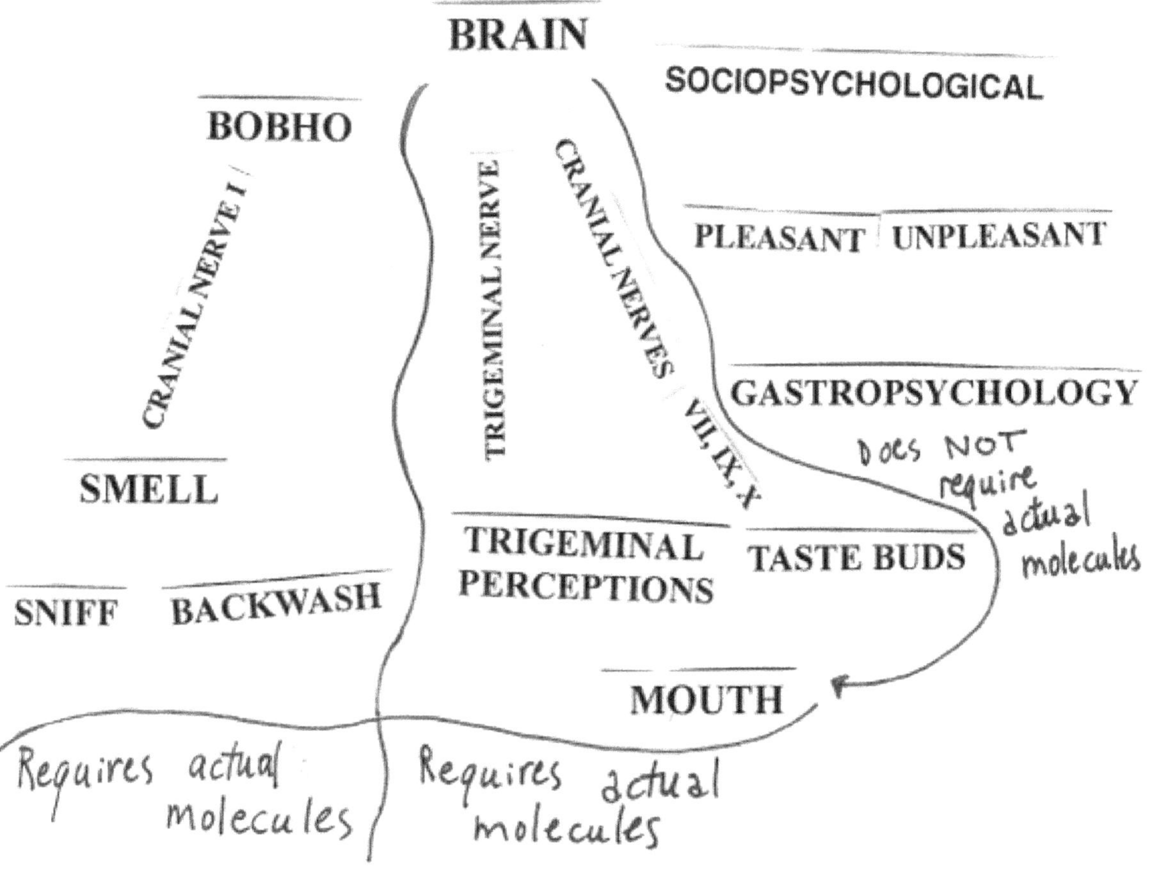

Figure 34. This flow graphic is trying to explain a very complex issue and is in stark contrast to the usual "smell + taste = flavor" simplicity. It is totally original so can't be compared to already-published versions.

EMOTION EMOTION EMOTION EMOTION EMOTION

DECISION TO INJEST

GASTRO-PSYCHOLOGY (*Gate 1*)

SMELL

(*sniff*)

TASTE
(*Tongue = sweet, salt, sour, bitter*)

TRIGEMINAL PERCEPTION

(*texture, temperature, pressure, pain etc*)

DECISION TO SWALLOW
(*Backwash to nose*)

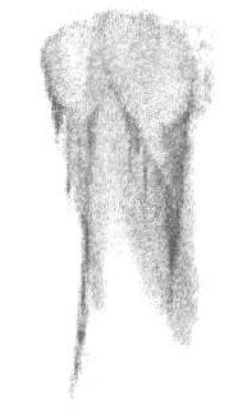

BRAIN ANALYSIS

(*Identify item: possibly distorted by emotion, trauma,
smell/taste disorders etc*)

Figure 34b

Brain Pointillism: Bring in the French Artists

What's my point?

Art can sometimes help explain science.

Put simply, BobHo takes a load of diverse data and instantly tries to make sense of them. His Majesty does not need all the data at once to make a decision. Here's a real-world example: I may arrange to meet my loved one on a railway station platform after dark. The lighting is muted crepuscule. A figure walks toward me. It's her! I don't need to look at her driving license to know. The way she walks may be enough. You have a mental picture pattern of how she walks. In World War II in Britain, they encouraged people to identify bomber planes from Germany just by a broad impression—no binoculars necessary. They called it "GIS"—general impression and shape. Since then, birdwatchers identify species in just that way. Likewise, BobHo fills in the blanks from memory to identify an item. You are offered a banana. BobHo instantly says, "Even from the look of that, I know what it tastes like because of my early childhood memories. It's a banana. I like bananas. Mmm." Job done. Sometimes a smell is enough, but sometimes BobHo needs a little more information from the trigeminal nerve—mouth information to fill in the blanks. "I think I know what it is. Er, wait—there it is. Crispy, crunchy potato chips." The idea of the brain processing data points/pixels and mapping them to make sense in the big picture is not new by a long shot. In 1886, the French artists George Seurat and Paul Signac developed an art technique as a branch of impressionism.

The miracle of Wikipedia expresses it this way:

Pointillism: a technique of painting in which small, distinct dots of color are applied in patterns to form an image.

Georges Seurat and Paul Signac developed the technique in 1886, branching from Impressionism. The term "Pointillism" was coined by art critics in the late 1880s to ridicule the works of these artists, and is now used without its earlier mocking connotation.[1] The movement Seurat began with this technique is known as Neo-impressionism. The Divisionists also used a similar technique of patterns to form images, though with larger cube-like brushstrokes.

Put in blunt terms, the idea may be represented this way: if you look closely at the data points (paint dots), they are just that—data points, or dots. However, when you stand back, you start to form distinct images. In other words, if you look too closely at something, you might not see it. I hope you can see where I am going with this. BobHo takes pieces of King Flavor information and maps them with the hope of being able to label them. Memory is vital here. You only know what a banana is after having eaten one and someone telling you, "You just ate a banana." In a real-world scenario, people in tropical regions only know blueberries after being exposed to a blueberry sample. Without this experience, someone from Uganda would certainly not be a good source of sensory advice on blueberries.

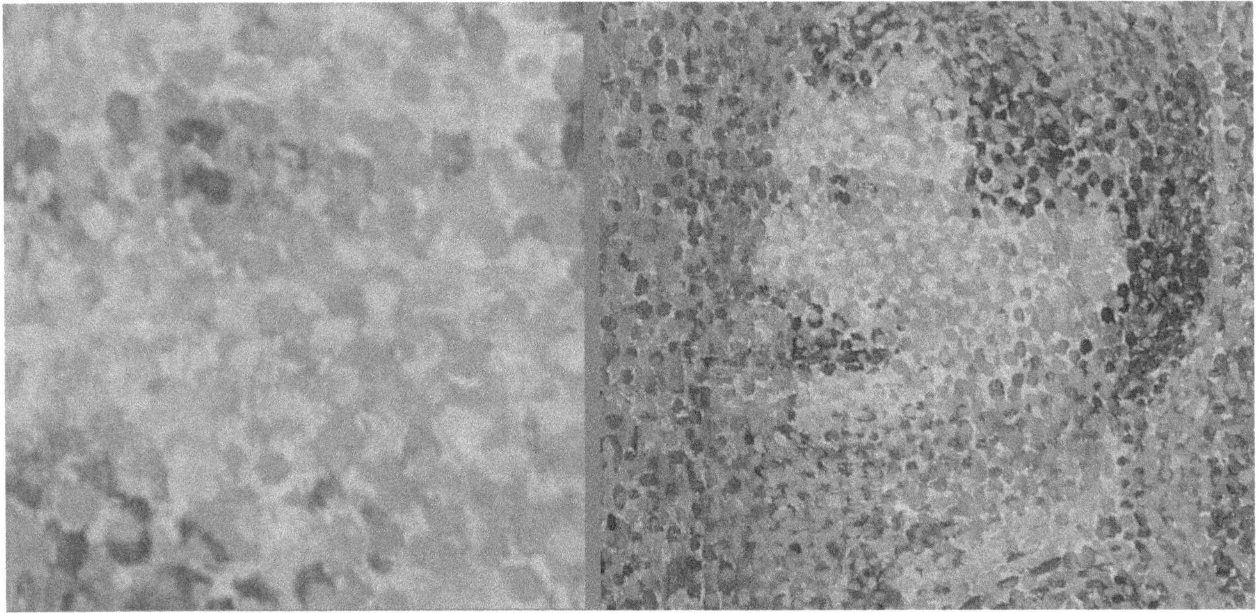

Figure 35. The left image is taken from the man's eye next to it.

In the left graphic depicted in figure 35 are individual data points, which are meaningless. Stand back, and the brain interprets them as a collective whole—in this case, an eye. It does the same with smell and taste and memory data points. I really wish that I had thought of this concept because it is terrific. In fact, I use it in my lectures, and it always receives rave reviews of understanding. How many times have I heard about someone smelling an old book or piece of clothing and the person saying, "That brings back memories of my grandmother's house"? The brain maps back to our earliest experiences, and time does not dilute them. Let me emphasize that: time does not dilute them.

The Difference Between Taste and Flavor

What's my point?

Confusion reigns everywhere.

Why is it such a problem to understand that taste and flavor are not the same thing? Even science journals, wine tasters, beer judges, and others who one would assume would know better are forever getting them muddled up. Flavor perception is such a confused, messy muddle as it is without getting two simple words misused. Perhaps the mix-up is the muddling of science and linguistic definitions.

This just in, as an example of taste and flavor confusion: Sam Adams brewery hosted a beer tasting on a plane. Quote: "Hosting an in-air tasting isn't exactly the best way to show off a flight"— I hope they knew that was a funny word choice— "of beers. A study by the Fraunhofer Institute for Building Physics in Germany found that taste perception is decreased by around 30 percent when you're in the air." Wrong, wrong, wrong. I think they meant "flavor," not "taste perception." Even on the beloved and trusted National Public Radio's program *Science Friday*, a correspondent mentioned a sour smell. No, no, no. They meant the smell of something which from their personal memory was sour tasting. Regarding the meaning of taste and flavor, it is not hard to find sources who should know better using them interchangeably and wrongly.

Taste is a verb and a noun and an adjective, if you include "tasty" as in the title of the book *Tasty* by John McQuaid. Did anyone ever say, "I am fascinated to taste tasty tastes"? I just invented that, but you get my point, I hope—or not. You would not be shocked or disturbed to hear someone say, "This cheesecake has a wonderful taste." The English is correct, but the use of the word "taste" is not. "Wonderful flavor" is what was required. I am fully aware that, day to day, none of this would matter, but with enough confusion residing in the smell, taste, and flavor worlds, it is my duty to be as exact and pedantic as I can in this book, even if some find me annoying. After all, I did attend Colchester Royal Grammar School.

To further disentangle yourselves from the meaning cobwebs, you will have to finish the taste and flavor sections of this book. Smell plus taste produces flavor. Wrong, wrong, wrong, despite how the major English dictionaries define it. Let me continue this exploration of taste with a disclaimer (okay, I admit, I like disclaimers). The same goes for all the other chapters too. This book is not the definitive book on all things flavor. It is a guide to make you think, question, and distrust (including myself), and I knowingly launch into generalizations to get a point across. Let me put it this way: if you want to know about all the nerve pathways concerned with taste, then go ahead and read all the books out there that specialize in that sort of thing. Just don't expect to make friends by discussing your findings with them. When I talk about trigeminal nerves and their function, I really do realize that they do not work in isolation. I am just trying to make it accessible (not simple) information for you. Don't go knocking on my door at three o'clock in the morning—or even worse, reviewing this book at any hour of the day—with criticisms that I got some details wrong. It would be no news to me. I am deliberately getting things wrong. It is the concept that matters. Now, leave me alone, and carry on reading.

I may use "taste buds" when referring to the nubbins on the tongue, but essentially, I could just as easily say "taste receptors." Taste buds house the taste receptors. Traditionally—and by that, I mean in just about every other book on flavor perception—taste is married to smell to produce flavor. Referring yet again to King Flavor at the beginning of this chapter, you will remember (because you did not skip past it) that I have relegated taste to a whole bucketful of other mouth stuff. I must find a better name than "mouth stuff," but that's what it is for now. I take that back. How about I call it "mouth trigeminal perceptions"? So, in less than an inch, I go from idiot to Mr. Neurological Smarty-Pants. See chapter 6 for the full exposé.

The word *taste* is always causing confusion because it can be used in so many situations. Earlier on, I mentioned, "I like to taste tasty tastes." Broadly, taste gets bandied about like a universal flavor hook. If it goes in your mouth, it must be taste. That is where, apparently, we taste things, after all. It's like when the word "like" is used out of true context like a modern phenomenon: "This pizza is like, really tasty." Wrong on two counts. "This pizza is really flavorful." That is, of course, if the pizza in fact pleases you. I once had a pizza that didn't taste as good as the box it was in. It was at the Great International Beer, Cider, Mead, and Sake Competition, for which I was chief steward in charge of thousands of bottles, all full. I had plenty to distract me, but I really did say that about the pizza.

At the beginning of the book in the "Apologies and Disclaimers" section, I have confessed to being picky, particular, persnickety—pedantic, in other words. An unashamed, self-confessed pedant am I. This is my excuse for unrelentingly hammering home the difference between taste and flavor. Flavor is king. Taste is the servant of flavor. Taste works in the mouth department, which is a busy place with all the other aspects to deal with, which include pain, temperature, texture, astringency, pressure, juiciness, carbonation, pungency, spiciness, and—wait for it—tastes.

Taste is not the big shot it thinks it is. You can take your seat now, Mr. Taste.

TASTE BUDS: FUN FACTS

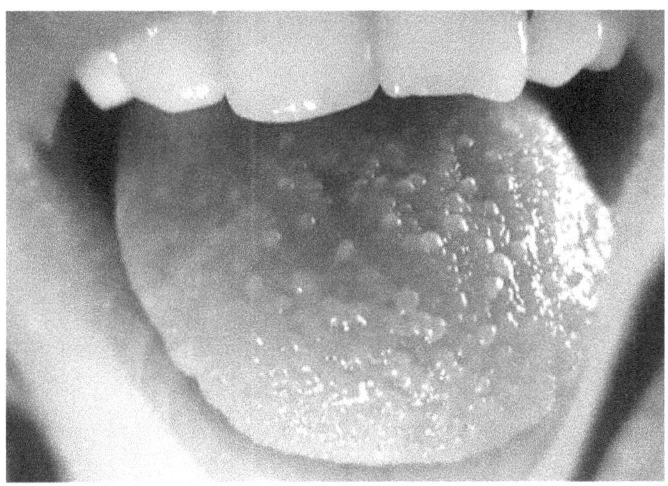

Figure 36. What are those nubbins on the tongue for?

Are lumpy tongues just a distraction from bad (British) teeth?

What's my point?

I hear you say that you are still wanting me to dig deeper, with the hope that I will bury myself. Let's try then. It still amazes me that regular foodies, drinkies, and even the professionals among them broadly believe that taste buds detect flavor. In that case, what are they for? "To tell you what things are." When it comes to taste buds' work description, this is a very short manuscript. Considering that taste buds are dotted around the buccal cavity (mouth), with the tongue being the epicenter, you'd think that the average intelligent person would have a good idea as to what their function is. The fact is that their job description can be summarized as this: to detect sweet, sour, salt, and bitter substances. Pretty basic, eh? As you know by now, I don't accept umami, and if you'll just trust me and settle down, I'll explain in depth why in a few pages' time.

I believe taste has its roots in evolution so that we don't kill ourselves inadvertently. Even then, taste buds are really restricted in their talents. They detect sour but are clueless as to the type. They know not the difference between a lemon or spoiled milk. And they don't care. Their job is to stop us consuming something that might make us ill or kill us. Dolphins are apparently intelligent and don't like to kill themselves either, but simply detecting salt won't help them for obvious reasons. They don't eat sweet things, which is why perhaps they always look so sleek and healthy and tend not to hang around the local chocolaterie. In fact, they don't need taste buds at all, which is maybe why they don't have them. I can see Darwin smiling in his grave. Pretend that you are a dolphin who is not owned by the Disney corporation. You are out there in the open ocean and see a silver fish—a nice-sized one. It sees you and swims rapidly away. What more do you need to know before chasing after it, grabbing it, and swallowing it in one gulp? It must be a healthy snack—it swam away at all speed. The dolphin had no need to sniff it. Just chase and swallow it whole. No need to chew as your mother told you—or not. What did she know about dolphins? My mother served British Queen Elizabeth II on occasions, and this subject never arose, not just because one is not allowed to address Her Majesty. Don't speak unless spoken to.

So, if taste buds do so little, why should the brain care? The buds send their information to the brain via cranial nerves. The brain receives smell-specific data via the olfactory bulb (BobHo) and mixes them with other data that it is receiving from the tasty item in question. The brain tries to make sense of it all by accessing information from memory and emotion. The taste bud data mixes with the ingested item's other data to make sense. For example, sour plus memory might mean lemons. It maps flavor so that the brain may identify it. The tongue papillae (nubbins) enclose the taste buds. As previously indicated, there are around 250 taste buds on each papilla. The taste buds have receptors and mechanisms to detect sweet, salt, sour, and bitter.

The four tastes are not just different in name. The bitter receptor types far outnumber the sweet receptor types—currently, two for sweet and twenty-four for bitter, but that will probably change tomorrow morning, science being science and operating as it should. Putting it another way, bitter perception is much more complicated than sweet perception. Sour is probably just a pH issue. This next statement, I think, is rather stunning: the taste buds don't give a hoot about the source of the sweet or sour or bitter. Lemon and lime, both are fine. Sweet is sweet. Bitter is bitter. Sour is sour. For example, the tongue knows that a citric liquid is sour but does not know lemon from grapefruit from orange or a bucket of worms—the same as the wisdom of wine sommeliers in the real world. Ouch! That hurt. It is also what my wine importer friend says. If you temporarily lose your sense of smell because of COVID-19, you can *still* taste and perceive flavor. The taste nerve pathways don't go through, or even near, the nose. However, losing your sense of smell will affect King Flavor, even though you can still taste. Yes, confusing, which is why doctors (reluctantly) engage me to explain it to them. Doctors love being educated by GED nurses and accoucheurs.

Hold your horses, cowboy; it's 2024, for goodness' sake. Is this where we get the boxing gloves out? "In the blue corner, we have four tastes. In the red corner, we have five or more." You have no reason to be aware of the debate on how many tastes there are. Cats don't care; they can't taste sweet (they like the fat in milk though). Dogs don't care; they can't taste salt. Dogs can taste cats though; yummy. Whales could care even less because they just don't need to detect much beyond salt. Dolphins agree with them. Octopuses don't. Humans, though, do care. Confused? Most books and authorities are in the five-taste red corner. I suspect that is not because they have actually thought about it but is more that they find it easier to just copy other people. To break the tension, I'll confirm what you already probably knew: most espouse sweet, sour, bitter, salt, and umami. Damn that umami. Now you see why I am a taste pedant. We are not even in agreement about how many tastes there are. To me, and others: aren't you proud of me for mentioning evolution in the previous paragraph?

Umami and Monosodium Glutamate (MSG)

Let's get my taste nemesis addressed right now. Feel free to disagree with me, but please base your opinions on science, not flimflam. So, what is the problem with MSG, which provides the umami effect? The trouble started with an uninformed doctor who was not very big on science. He went to a Chinese restaurant and got sick. He did not blame being unable to read the menu in Chinese. What caused it? The umami MSG, of course. There is no other explanation, is there? He then published his emotional, nonscientific whining in the *New England Journal of Medicine*. Ooh, scary medical journal, so must be true. With a name like "monosodium glutamate," it also must be some dangerous chemical concocted by those crafty Chinese. Wrong—you probably consume it almost every day in most of your food without even having to pick up the MSG saltshaker. It is, after all, a salt alternative, even though it has no taste in itself. It enhances tastes. It's in most plants. There is, as I write, absolutely no proof that it makes people sick. Why should it? It is naturally occurring. Naturally occurring is not always good. Let's not discuss poisonous mushrooms or arsenic, Gerry. It may be a nocebo. You don't know what a nocebo is? Surprise, surprise—it is the opposite of placebo. "Placebo" means "I please." Nocebo means...you've guessed it.

Homeopathic medicines contain no active ingredients, but they please enough people to make them a billion-dollar industry (flimflam beyond placebo?). By the way, I used to practice homeopathy professionally in South Africa, so I have experience. Many of my patients were thrilled with my care, although I did not cure any infectious diseases or cancer. If you believe that homeopathic medicines help you, great. That's what placebo means. Just think twice before you depend on them for serious diseases like tuberculosis, syphilis, or cancer. An extreme example of the nocebo effect is when someone has a curse placed upon them and they respond by getting ill or worse. It has happened. Don't tell me you have never had "the finger" pointed at you.

MSG is a nocebo. Just the word can displease people. That's why some restaurants have a notice saying, "NO MSG USED." Now, the educated among you are going to tell me that glutamate receptors have been identified on the tongue and in the gut. Correct. Just because they are on your tongue does not mean that they are active, though. Men have nipples, but that does not mean—you get the idea. Glutamates in food are probably in their "bound" (nonactive) state. Just sprinkle your MSG to brighten your meals and stop worrying. Not too much though—moderation in everything (except moderation). Other tastes that have been proposed include fattiness (oleogustus), temperature, texture, astringency, pressure, juiciness, carbonation, pungency, spiciness and—wait for it—tastes. I don't understand either.

Taste and smell are always the servants of King Flavor.

Can you have taste without flavor? Can you have flavor without taste?

Speak up, I can't hear you! Yes, you can have taste without flavor.

Let's keep it somewhat simple. It would make sense if, on an evolutionary level, tastes labeled nutrients for homeostasis. Since last year, when I wrote that there is a move to rename bitter receptors as nutrient receptors. Tastes must be in solution on the tongue to be detected. Not so with smell olfactant molecules up the nose.

Salt: useful to us on a cellular level. We like mild saltiness. Salt dampens bitterness. Overly salty food makes me cough. It doesn't make me a bad person, though. I don't understand either.

Sweet: we are born enjoying it the moment we breathe the outside air. We start life (historically) consuming sweet mothers' milk—baby's only source of usable energy.

Sour: without sour perception, we may inadvertently kill ourselves. It is useful to identify spoiled or unripe food. We must learn to enjoy it though—hence, sour beer lovers generally won't admit to enjoying their first one.

Bitter: without bitter perception, we may die. Potentially lifesaving when identifying poisons. Put that poison ivy down. Don't eat it. Bitterness triggers relaxation of the lungs to keep the airways open. This may come in useful if you are in anaphylactic shock and are on the verge of croaking. We are born disliking tart vegetables and IPA beers. Learning can be fun though, after the initial aversion.

So far it has all been about the Bee Gees song "Stayin' Alive." Current thinking is that bitter receptors throughout our bodies also detect nutrients.

Next? There is a serious move to include fat as a valid taste—oleogustus. I see their reasoning but come to an opposite conclusion. Would you believe that sweet and fat rarely occur together in nature? Mothers' milk is an exception. Tubs of ice cream don't occur in nature.

Umami: What an Absolute Kerfuffle

In 1908, a Japanese gentleman named Mr. Ikeda (who had nothing to do with the Swedish home decorating stores) told the world that glutamic acid in kombu seaweed supplied its palatability. Glutamates, the salts of glutamic acid, are present in most living things. Where would mothers' milk be without them? Are you still awake? The world might not have cared, except that the next year, he patented monosodium glutamate (MSG) as a flavor enhancer. Did he do it for money or simply to amuse me? MSG occurs naturally in common foods like tomatoes, cheese, meat, mushrooms, and lots in breast milk. It is entirely possible that if he had not patented it, I would not be writing this chapter now, and the world at large would fail to care.

Some professional sensory scientists see the umami debate as hinging on whether it is a con or not, because so much money is involved. I will try to tackle this subject from a skeptic's standpoint—just the facts.

Umami taste has been described as deliciousness—savory. Sampling sweet, salt, sour, and bitter, I recognize them immediately. If you ask me to give you some umami food, I will have little idea what to do. Umami is described as deliciousness. A piece of milk chocolate is deliciousness to me, but I don't think it is umami. At the end of this chapter, I cover umami again with details of a test I did on flavor professionals with umami paste. One hundred percent umami. That's right; none of them got it. Something else disturbs me too. With the other four tastes, there is an evolutionary homeostasis advantage to being able to identify them, but if I can't identify umami, how does that prevent me from living a long and prosperous life? It is so dissatisfying to write about umami. Permit me to ask you to make your own mind up as to if and how it exists. By the way, being a skeptic, may I say that if umami is a taste, this will help MSG sales? It's not just me saying this. A doyen of taste is Linda Bartoshuk, and I give her credit for bringing this up, although I disagree with her on other subjects like supertasters. Why is there always a "but" with you, Gerry? In a nutshell, here's how monosodium glutamate operates in the body.

A food item on its own might be perceived as boring. The same food item with MSG...whee, delicious. The word *glutamate* derives from an amino acid, which is an essential building block for proteins. When you eat a piece of meat, for example, you are eating protein, which contains amino acids, including glutamates. Pure protein has no taste or smell. Compare raw, straight-out-of-the-cow beef to the same thing cooked. Amino acids are bound up chemically in the protein. To split them, in order for them to enter the blood stream and be of use, one needs to digest them. This does not happen in the mouth but lower in the gut. Protein molecules are too big to be identified in the mouth. We have glutamate receptors lower in the gut. Now this makes sense. The gut detects glutamates and sends signals further up the neurological chain, indicating that some high-value nutrients have arrived. You decide. To emphasize, the gut detects glutamates, but the mouth doesn't. So how are we supposed to detect umami in the mouth? Perhaps the gut can confer with the mouth and brain and say, "That was a piece of meat and a high-value food, so I should try to eat that again in the future." A piece of cardboard with the word "glutamate" written on it won't do that. This does not explain why we would want glutamate receptors on the tongue. It all comes down to receptor sites for umami, which exist, but do they do anything? Why would we want them on the tongue?

What about other creatures? They perceive smells and tastes, but not in the same way as we humans. Why not? How about evolution and homeostasis? Yeast cells identify carbohydrates in order to ferment them and provide alcohol in beer and wine. Dogs need to smell but not necessarily taste. Smells help them find food and potential mates in heat. Dolphins are not discrete eaters. If it is silver and wiggles away fast, it is food. Grab it and swallow it whole. There is no equivalent of a wine taster in the animal world. When a whale swallows a shoal of fish or squid, they hardly touch the sides of the gullet, or whatever whales have. No intellectual reflection or delicate chewing or spitting out the old ones for them. No "delicate notes of this and that." No discussion on fish scale texture. The giant pacific octopus is another creature entirely. Octopuses are very choosy. They make our ten thousand taste buds look pathetic. They have ten thousand on each of their 1,600 arm suckers. That's a cool sixteen million buds. Mind you, that's typical of octopuses—enough is never enough for them. They also have three hearts and eight arms. Do you? The point is, why taste if you don't have to? On an evolutionary level, it is a moot point. And so, I play it safe and conservative and discount umami in this book. Honestly, we can do very well without it. Feel free to prove me wrong, as always and allow me to eat my mushroom without thinking about it.

Taste Receptor Functions Beyond the Mouth

It still comes as a shock and surprise to most people when they hear that taste receptors exist all over our bodies. I once gave a lecture to doctors, and before the talk, I asked some general surgeons, "You know that we have taste receptors throughout the gut, don't you?" They replied, "We do?" Now you don't feel so bad, do you? We have sweet receptors in large numbers on the tongue, but they extend down the gut. Why would we need to perceive sweet down in the small intestine? Of course, that old devil, homeostasis. Homeostasis is all about keeping us alive and well. The gut receptors monitor sugar absorption. They are not concerned with informing us about the presence of a sweet substance. Sugar absorption influences blood glucose. Blood glucose is regulated by insulin, which is tied to diabetes and other conditions. Time to study insulin resistance and metabolic syndrome if you want a real downer. Other receptors regulate other nutrients, but sweet may be the most stunning and easily understood example. These sweet receptors are also triggered by artificial sweeteners, and perhaps this is why artificial sweeteners may not lead to weight loss. The body thinks it is receiving energy and nutrients. It is being lied to. That has no place in homeostasis. At present, I only see negative effects from artificial sweeteners. "Nothing for nothing," as they say.

Testicular Taste Receptors: Not Just an Attention-Grabber

Humans routinely use mice and rats for medical research because they are genetically like us in many respects. That does not include fur and tails. For one research project, mice were genetically altered to edit out bitter receptors in their testicles, and these mice were encouraged to multiply by simple breeding. The mouse cage becomes a fun house. Use your imagination, but no way, José. It seems that bitter receptors in the testicles are linked with the viability of sperm. I can't believe your mother never told you that. Having no testicular bitter receptors doesn't augur well for those who plan on a big family, as mice are wont to do. That is not something I have to concern myself with. It still comes as a shock and surprise to most people when they hear that taste receptors exist all over our bodies. That includes your brain and heart. Bacterial infections often release bitter compounds. If you are a supertaster by definition of you having lots of bitter receptors, does this mean that if you get, for example, a sinus infection, your body will respond quicker and fuller? How about, yes? Good old evolution—always looking out for us.

Sweetness: Good, Bad, and Rats

Fine, I'll discuss rats first. Do I know you or what?

Figure 37. *"I have no complaints regarding the food, but why do they insist on drawing our blood samples afterward? I smell a rat!"*

Scientists experiment on rats because, in many ways, they are like us. Even mafia celebrities apparently say, "You dirty rat." Given a choice, rats prefer sugar to cocaine. Where am I going with this? I'll tell you. Do you want to catch a rat? Use saccharine, not sugar. Rats prefer it. Better than cocaine, even for them. But what do they know? You are aware, I presume, that sweetness comes in many forms, and it should come as no surprise that our love for sweetness harks back, on an evolutionary level, to survival and our need to consume high-energy foods. In the early days, this was frequently honey and in-season fruits like figs and dates. Sweetness is universally enjoyed, but not at extreme levels. Now, if you want me to define extreme, I can't. If Jim defines extreme as gagging, cloying, sickly, syrupy saccharine, this level may be pleasant to Linda. In fact, Linda Bartoshuk is a famous sensory scientist who admits to not knowing what too sweet is. I do; half a donut, and I'm done.

We have various taste receptors—sweet for sweet and bitter for bitter. That would seem logical, but nothing is logical with flavor perception. Sweet perception is affected by bitter receptors. You're confused? You are not alone. Some proteins taste sweet. Sugar activates bitter receptors. Can I go home now? What does that mean in the real world? Research is generally done on animals which are genetically close to us. Pets are physically close, but not genetically close. The commonest pets—cats and dogs—have different dietary requirements from us, although I agree that dogs are more like us than cats. A scientist in Ethiopia once told me in all seriousness that dogs make better pets than cats, because if you die, your dog will guard you, but a cat will eat you. I kid you not. Don't look to your cat for information, because she can't taste sweet, although your dog can. Before you rush out to the pet store to buy rodents to test because they are genetically quite close to us, understand that even they are not good at this taste stuff as applied to us. Rodents are known to have a sweet tooth, but neurologically, they perceive it differently from us. The takeaway? Never trust a rodent's opinion when it comes to sweet. Whales are even worse. I am trying to emphasize that we are not like other animals. To a large degree, you have no choice regarding what you like to eat and drink. Be gentle to your children. Yes, I realize that this is a massive subject and I have left out 97 percent of what I would really like to say.

Sweet: Straight Out of the Womb

What's my point?

Mum's milk has it all—for the time being.

Going through life, we can learn to like (and dislike) certain smells, tastes, and flavors. There is not even one universally liked or disliked flavor after breast milk. Believe it or not, chocolate is not liked by everyone. Then again, pure chocolate (cacao) is bitter, and not sweet. At the far other end of the scale, the rotting shark meat, *hákarl,* is enjoyed by many Icelandics and apparently a few other people who are less desperate. Sweet taste is hardwired at birth. Flavor perception is learned. Smell is learned too—just ask any perfumer or wine sommelier. Let's face it; we need some good sustenance after sliding down that slippery birth canal. It's been a bumpy nine months. Did you ever meet a newborn baby who turned her nose up at mothers' milk? On the other hand, did you ever meet a newborn who could not get enough of bitter Brussel sprouts purée?

The point of the foregoing is that we are obliged to learn to like or dislike everything we consume. However, straight out of the womb, we must rely on some sort of immediate instinct. BobHo knows what is good without any experience. The newborn does not have much experience of flavors, so decisions on drinking are potentially life and death matters. Literally. BoHo always says yes to mothers' milk. If the baby said no to mothers' milk, then of course that would mean...well, you can work it out. It's not pretty. Lots of sobbing grandparents.

However, what does the baby know at this stage of its life? Not much. Just a huge unconscious interest in homeostasis. After all that swimming then sliding down the chute, a baby needs a drink. Let me see, what shall she have? Something sweet to give her immediate energy so that she can cry and cry and cry and cry. Hold on; she can smell it. It's just above her at the nipple faucet. Mum should really get that checked out; it's leaky. Still, the smell made it easy for her to find. Lots of energy-giving sweetness plus more because pure sweetness has no smell. More milk please. She's not ready for sour, bitter, and salt yet. She does remember the flavors of some of the foods Mum consumed during her last trimester though. She especially liked the carrots and garlic. Yes, the pregnant and breastfeeding mother can influence taste preferences in the baby by what she herself consumes. Taste is hardwired at birth. Flavor perception is learned. Research it. I dare you.

To repeat from earlier, nothing smells sweet. Sweet is not a smell. Sweet is a taste. It is associated with sweet things though. Linguistically, we may say, "What a sweet little girl." Vanilla, cocoa, and cinnamon are all bitter, not sweet. However, we are used to having them combined with sugary products.

Children and Sweetness

Imagine you have twelve children, half of them adopted. Let's be clear: I am just saying "imagine it." Don't take it as an order. Firstly, why so many? And secondly, well done regarding the adopted ones. Just for fun, line them all up at the kitchen counter and offer them a lemonade. Then you give them a chance to add more sugar, should they wish. Some probably will, and some won't. The genetically similar ones (yours) will tend to be more similar in their taste preferences than the genetic interloper ones. Genetic inheritance is at play. If your child Angel says something is too sweet, that is useless information, because adopted Vladimir might not agree.

You can't smell sweet. How many times have I got to repeat that? I'll keep repeating it because people who should know better continue to say it. Can I interject here that I realize sugar is addictive to many? It lights up the same pleasure brain pathway as alcohol and perhaps cocaine. I'll do you a favor by stopping just there. I can't help myself; why do people acquire alcohol and sugar addictions but not broccoli addiction?

Figure 38. "Mary, now there's something you don't see very often."

I am politely requesting some critical thinking from you here, please. I wrote to the editor of *New Scientist* journal to tell him that, in reference to an article that mentioned a sweet smell, sweet is a taste, not a smell. We can't smell sweet. He responded that even Shakespeare mentioned the sweet smell of success. I responded that I thought he was editor of a science, not literature, journal. That did not elicit a laugh or even a response. If a chef says something smells sweet, smack her around the ear with a limp banana skin. I'm joking; only do that if the chef is not a man who also has no tattoos. It goes to say that nothing smells bitter, sour, or salty. Let me state this: sugar intensifies flavor. King Flavor. The whole caboodle. What is a caboodle?

Well, here is my answer. If you want to make grape juice taste grapier, you don't have to add more (expensive) grape juice; you can add cheap sugar instead. Why didn't the food industry think of that? They did, of course. Just read the labels and look under "added sugar." Children were tested, and it turned out to everyone's shock that the tubbies desired sugar less than the skinnier ones. Fatties were easier to satisfy with sugar. It was also found that it is the bitter receptor genes (not the sweet) that determined our sweetness preferences.

"Jim, I am worried that Angel eats too much sweet stuff."

"Penelope my love, I am a man and therefore know everything. The answer is simple: if we restrict her sugar and sweet intake, she will adapt to it and require less to satisfy her sweet tooth…if they haven't already fallen out rotten. Just joking, darling."

"Jim, I was reading the latest research from the Monell Chemical Senses Center, and they say it does not work that way."

"What? Are you going to tell me I am wrong?"

"No, dear; Monell is going to tell you that you are wrong. If you contact them, they may tell you that you are confused with salty, on which one can alter one's dependence."

"Let's be clear here: if we gave Angel no sugary treats for fifteen years, she would still retain her sweet tooth?"

"Oh, my darling, you are so quick."

It should come as no surprise that there currently exists an obesity epidemic among kids (and their parents). The whole subject is enormously complex, and I will not do it justice to address it briefly, as I am obliged to do. As I write, kids are born hardwired with their sweetness preferences. Trying to limit their sugar intake does not dumb down their desire, as just discussed. Preventing them from eating candies will not dull their desire to eat candies. Strangely, though, if there is a preference for lots of added salt on food, this can be dulled by reducing added salt. One can desire less after a period of withholding. On a purely personal note, salty food makes me cough, as previously mentioned. Why? I have no clue. Monell Chemical Senses Center tells us that "our perception of sweetness is inborn in our genetic makeup." In other words, sweetness levels mean different things to different people, as I have stated ad nauseam.

Too Much of a Good Thing?

Can you have too much of a good thing regarding sweetness or anything else? Not wishing to get morbid, but with addictive drugs like heroin, more is never enough. Satiety does not exist. There is always room for more. That is why I maintain that caffeine is not addictive, because there comes a time when enough is enough—generally before the thirty-sixth cup before lunch. It also explains the Thanksgiving-meal phenomenon whereby there is always room for a little more. You are permitted to prefer very different definitions. There was a lovely study done where they tested self-professed chocoholics. Essentially, they each received a square of chocolate every five minutes while their brains were scanned with an fMRI scanner (see fMRI under "Oddities" in chapter 7). Initially, the happy, happy parts lit up, but after around fifteen squares, the scans went from happy, happy to throw up, throw up. Interestingly, revulsion came after very different numbers of chocolate squares. Some volunteers required twice as much as others.

"Too sweet" is a very personal thing. The fructose half of table sugar is what the tongue loves, so the tongue must also love high fructose corn syrup, right? The glucose half is not very sweet, although every cell in the body can use glucose, but not one single cell can use fructose.

Figure 39.

Consumers apparently notice the difference, and there are other horrible negative health issues which have cast it in a very dim health light. Obesity is one of the mildest undesirable conditions associated with it. Obesity is not the problem; it is the result of the problem (metabolic syndrome). Humans just love sucrose, which is half fructose. Try telling the liver that fructose is not alcohol. More accurately, it treats it roughly the same as alcohol. Tell me about fatty liver disease again. We have entered the realm of discussing the concept of addiction—a big, variable subject without a simple definition.

Nonnutritive Sweeteners (NNS)

Lose weight? No need for me to. If you think taking artificial sweeteners will help you lose weight, then think again. They won't. They may even make you experience hunger and eat more and put on weight. "But they contain no calories, Gerry." Tell that to your tongue, brain, and gut. The principle of NNS is to flat-out lie to the brain. Just eat the damned sugar if you must. Better still, don't. With NNS, you are messing with substances that nobody properly understands. Ask Harvard researchers. Go on. Ingesting NNS means you are messing with hormones that you have not evolved to utilize in this way. Hormones? Think insulin hormones and diabetes and heart disease and obesity. There are many sugar alternatives, which are described variously as low-calorie sweeteners, nonnutritive sweeteners, and artificial sweeteners. Just picture those little packets in the jar on your table at a restaurant. They don't help diabetics or aid weight loss. They include Sweet'N Low (saccharin), Equal and Nutrasweet (aspartame), and Splenda (sucralose). Their flavor will not rock our universal boats and so only account for a relatively small percentage of the sweetener market. Artificial sweeteners also trigger the release of as much or more of the hormone insulin as real sugar (yes, it tricks the brain) and negatively affects the gut microbiome, therefore negatively affecting homeostasis and producing bloating, gas, and generally disturbing the good gut bacteria. Wonderful, is it not? This is where other authors would insert a reference; not me, because there are so many being produced as I write this. I am not just being lazy, I promise. They don't even help you lose weight, according to the latest research. Speaking simplistically, insulin aids fat storage. Is this where I talk about exercise? Which exercise? How much? Lactate levels? Mitophagy? You should stop me here—it's a pity we ever met.

The Ultimate Sweetness

Sucrose sugar has been described as a mitochondrial poison. (Does "pure white and deadly" ring a bell?) All cells have mitochondria, so that means...?

Table sugar, which is mostly simply referred to as "sugar" (or sucrose), chemically has two parts: fructose and glucose. Every cell in the body can utilize glucose, yet not one cell can use fructose for energy. When sucrose hits the tongue, it is an almost perfect fit for sweet receptors. No distractions. No bitterness. Just lovely, lovely irresistible sweetness that allows us to concentrate on the chocolate flavor attached—if that is what we are eating. It is the fructose (fruit sugar) molecule that provides most of the sweetness and glucose only to a lesser extent. All our cells can use glucose. None of our cells can utilize fructose. There, I said it again. (Refer to the figure 39 cartoon.) That lovely bottle of clear apple juice should be regarded as a bottle of potential liver fat—a bottle of alcohol. But hold on, Mr. Spoiler; this juice is 100 percent organic. All right, have it your way—100 percent organic liver fat. There are, however, things sweeter than sucrose. What does that mean in practical terms? Well, if you wanted something containing as much sweetness as twenty thousand teaspoons of sugar, then you could use just one teaspoon of Advantame artificial sweetener. Think of all that weight loss! That's perfect, then; so, we will all just use Advantame for everything. The trouble is that we simply don't enjoy Advantame. What tastes just like the sugar that we love so much? Cane and beet sugar. Nonnutritive artificial sweeteners have drawbacks. Many leave a nasty taste in the mouth in addition to the intense sweetness. So surely "natural" stevia is the answer. It is natural and sweet. Ooh, organic must be even better. Some people feel an uncomfortable full-ness or nausea from it. Humans prefer sucrose. If we replace sucrose with low-calorie sweeteners, then all is well with the world, wouldn't you think? The full answer requires another book, but in essence, nonnutritive sweetener weight loss is just a myth. Fructose is what the tongue loves, so it must be crazy in love with high fructose corn syrup. It was developed because it is cheap to make and has a long shelf life. It is not about you. It's a pity that the tongue doesn't know much about the subject. (I'm saying that tongue in cheek.) Don't forget that the tongue does not know one sweet from another sweet.

Artificial sweeteners trick the tongue, but they don't trick the rest of the body. Put sugar in your mouth, and the message is sent out, "Body alert, sugar coming. Brain to pancreas: start squirting out more insulin." Ah, but these sweeteners have no- or low-calorie loads. They do not supply energy. However, research has shown that some hours later, you get the munchies and make up for that every time. Have a diet cola for breakfast, and by lunch, you have made up for it with other junk. So why not just drink more diet sodas? Because at some point, fat cells take them up and make you viscerally fatter. I say viscerally (organs) because fat under the skin is normal in preparation for starvation. No need to stare at her arm fat, stare at her beer belly. Sweet is sweet, but the artificial ones often stimulate bitter receptors as well. They may also just not sit well with consumers. We want our sugar.

It's soapbox time, folks. On a social note, the harvesting of sugar cane by field workers is horrible, physical drudgery and a very, very filthy process, which is just one reason you personally probably do not engage in it. The stubble is burned, which means the workers spend their days in smoke. After a day in the fields, the workers appear black, even if they did not start that color. Workers, even in Florida, get horrible lung disease, notably and tragically called canecutter's disease. They don't expect help because the sugar companies make huge financial and political contributions worldwide. In Dominica, a huge source of sugar for America, the workers are abused, and their families subjected to child labor and human trafficking. So next time you sprinkle that pure, white crystal on your food, think of them in black terms. When you read "Dominic" on the sugar package, know that the crystals may be pure, but the processes that produced them are impure. So, sue me. Sugar cane is a monoculture that supports very little in the way of wildlife except cane rats. When we consume pure, white sugar, it is at a filthy societal price, in other words. Even India has a huge societal problem with sugar cane harvesting for Pepsi and other corporations (sorry, Atul). Yes, I love my soapbox. So why don't we ditch all these other sweeteners and just consume cheap and easy-to-use sugar? Calories. Damn those misunderstood calories.

Sugar Beet

Where does your pure, white, crystalline supermarket sugar come from? Well, much of it is from these little beauties (I don't mean the man and wife):

Figure 40. *"Bill, just look at this little beauty."*

It is not unreasonable to assume that when most folk think of sugar, they think of sugar cane grown in tropical regions. Shock of shocks—around half of sugar sold in US stores and countries worldwide comes from the big, brown, lumpy roots shown above and below: sugar beet. It is loaded with sucrose and after processing is generally sold as something like "all-natural, 100 percent granulated sugar." For some reason, "cane sugar" sounds and sells better than "sugar from lumpy, brown roots," so they do their best to disguise the source without lying. Cane and beet sugar are chemically identical. Which would you rather your sugar came from: figure 41 or figure 42?

Figure 41

Figure 42

Stevia

What's not to like? All natural.

Figure 43

Take those all-natural leaves and just pass them through this all-unnatural chemical works, and you have white sweetener, sort of all-natural...apart from the massive chemical processing.

Figure 44

So why bother? Stevia is popularly sold under the trade name "Splenda"—no calories and two hundred times (give or take) sweeter than sucrose. Possibly a great choice for diabetics, although that can be debated. It contains at least twenty sweet compounds, none of which can hold a candle to sucrose. In front of me is a container of 100 percent natural stevia extract—natural stevia, not natural processing. It is heavily cut with lactose milk sugar. If stevia is so great, why add the milk sugar? Sampling it, I tasted distinct sweetness, but it sure is not sucrose. Bitterness comes to mind, which is not surprising when one considers that it contains chemicals, which bind to bitter receptors. You wouldn't get sucrose doing a thing like that.

Honey

So, what does honey contain these days? Surprise, surprise; some of the store-bought jars contain some honey. Just don't expect all of it to be honey. Most store-bought honey has had some of the genuine bee-produced honey removed and replaced by something sweet that looks like honey but is a heck of a lot cheaper and easier to produce. If you can buy the real deal from a local apiarist, then you have this magical, healthy sweetener, wouldn't you think? Believe what you wish, but it is still fructose sugar, glucose sugar, maltose sugar, sucrose sugar, and higher sugars. So why are we led to believe that it is healthier than table sugar? All very cozy and hippy, except that even at its best, it's nearly all sugar, which your body is happy to turn into fat around your liver, much as it does with alcohol. On the plus side, honey is expensive to buy, and most people cannot afford to consume much at a time or use it as their primary source for sugary sweetness. Honeybees require a diverse habitat, so they are important pollinators for their ecosystem. They are also fascinating to watch with their little "saddlebags" of honey. My wife Christina likes to tickle them, but something tells me that I am getting off topic. You will not be shocked to hear that I have a sensory certificate in honey. Apparently, some people assume I even have a certificate for being annoying.

Sweetness as an Analgesic

This is a little off topic, but studies were done with sucrose, and it was found that sucrose acts as a painkiller, especially in the young. Is your baby teething? Perhaps reach for the sugar candy before the drugs. It may well work with mouth cold sores too. Why can't everything be simple and predictable?

What is a Taste Modifier?

Put simply, a taste modifier on the tongue alters the normal taste of what you would predict and expect. Even though I said "tongue," it can be perceived anywhere in the buccal cavity. Don't forget the mouth trigeminal perceptions.

Wine tasters are famous for talking about wine astringency. This is a physical process, which I describe in my astringency section in chapter 6. Wine and other flavor experts famously get astringency and bitterness horribly confused. Essentially, the astringent chemicals precipitate proteins and give the impression of dryness by lying to the brain. In the mouth, this involves tongue mucus. If you coat the tongue, then it is harder for this process to operate. Cover the tongue with fat or oil and astringency is harder to perceive. What fatty food is regularly consumed with wine? Cheese, of course. So, what does that tell you about wine and cheese tastings? Come on, stay with me here.

A prime example of taste modification is the brewing of beer. The process starts with producing maltose—sweet maltose. In order to knock back the cloying sweetness, the malt (wort) is boiled with bittering hops. The bitter modifies the sweet, and one ends up with palatable beer. Taken to the extreme, one can consume hoppy, bitter India Pale Ales, or on the milder spectrum, "regular" lagers like Budweiser or wheat beers. Bitterness broadly separates the two as regards taste (you weren't going to say "flavor," were you?).

Miraculin

This has a terrific sense of humor and is ready to trick you in a nice way. Do you want some fun time in the taste house? Get some miraculin fruit extract and coat your tongue with it. The miraculin has nothing much going for it as regards taste, but it likes to mess with other tastes. Now suck on a lemon. Oh, yummy; so sweet. Miraculin fruit modifies sour to taste sweet. The laugh is on you.

Thaumatin

This one is very confusing to me. It is a fruit from West Africa. Even though it is a protein, it is considered (although I never personally measured it) many times sweeter than sugar. They tell me that the sweetness builds slowly and often ends with an unpleasant licorice aftertaste. I'll let you try that one.

Sterubin

Bitterness is modified by sterubin but is of no real interest to most people. It is enough to know just that it exists—a bit like Alaska and dark matter.

Milk and Cream

For me, these are possibly the commonest modifiers but are generally not described as such. Sweet modifies bitter so the bitter coffee is not so bitter. Even as a supposed purist, being a Specialty Coffee Association presenter, I use cream in my coffee.

Carbonation

Fizzy, carbonated drinks—what's not to like? If you are a laboratory animal, everything. They won't drink fizzy offerings. But you probably like the fizz of those bubbles bursting in your mouth. The thrill is not from bursting bubbles, I am sorry to tell you. Don't believe me? Try drinking carbonated water in a hyperbaric chamber. Well, go on then. You do have one, don't you? If not, borrow your grandmother's. Physics won't allow bubbles to form, but it is still perceived as fizzy to you. Dissolved carbon dioxide is carbonic acid. This fizz also makes sweet things taste less sweet. Just imagine how much sugar is hidden in a typical sweetened fizzy drink.

Table Salt: Nothing is Simple

Even scientists don't know exactly the mechanism of how salt dissolves in water. You are almost certainly really disinterested in hearing that. Just imagine how many other uninteresting facts I have respectfully left out of this book. Salt (sodium chloride) gives life to food. It is CPR for bland food. MSG is CPR for salt. Salt (sodium chloride) inhibits bitter. That is why some people want to put salt on their grapefruit. In fact, salt does all sorts of things to modify taste perception and mouth trigeminal perceptions. Regarding the latter, salt was found to improve the perception of product thickness, enhance sweetness, mask metallic (which does not exist) or chemical off-notes, and round out overall flavor while improving flavor intensity. These effects are illustrated using soup as an example. When salt is added to a soup, not only does it increase the saltiness of that soup (who'd have thought that?) but it also increases other positive attributes, such as thickness, fullness, and overall balance. Bring in the MSG. Low-salt alternatives are made from potassium (not sodium) chloride, which does not have the bad health, blood pressure rap that table salt has. Potassium chloride is cheap too; in fact, you can buy a forty-pound bag for around seven dollars, because it is just road salt used to melt snow. This is the chemical that the Specialty Coffee Association uses to replicate metallic taste. Funny how low-salt users never complain of a metallic taste. The Specialty Coffee Association does not want to talk about it, even though I am a presenter for them—I tried. Salt is also a flavor enhancer for fermented foods such as sauerkraut and kimchi.

Himalayan Pink Salt

Let's have some flimflam fun. The subject? Himalayan pink salt. You may think, like most of us, that salt is just salt. Sodium chloride. Perhaps a salt with added iodine is better. If, however, you desire the crème de la crème of salt, which contains no cream but lots of other magical health ingredients, you may be considering Himalayan pink salt. Asian. Inarguably must be exotic and organic too. Himalayan pink salt comes from where? The Himalayan mountains? Good guess, but no chapati. Himalayan pink salt comes from a part of Pakistan which is at least 190 miles from the mountains—a four-day walk for a mountaineer, and a five-day walk for me. Still, "Himalayan" sounds good. One advertiser says it is the purest form of salt available. Personally, I would have thought the purest form would be 100 percent sodium chloride produced in a laboratory or a big factory, which gets rid of all those "impurities" which make healthy salts so desirable. This is getting very silly, isn't it? I can hear Penn & Teller laughing.

All table salts are 90-odd-percent sodium chloride. Sea salt contains some impurities, which you can also get from a cheap mineral supplement. May I posit that to get a therapeutic dose of these trace elements, one would have to eat a rickshaw-load of salt? That may not be good for your general health in other respects though.

Red wine contains resveratrol, which may be heart healthy, but how many glasses or bottles of wine a day does one need to drink to get a therapeutic dose? Now we seem to have introduced another problem. Kosher salt is very lumpy and stimulates the trigeminal nerve. It is the same mechanism that potato chips engage to stimulate the trigeminal nerve with their crispy texture. Himalayan pink salt is great because it contains lots and lots of sodium chloride, just like…table salt. Magically, Himalayan salt does amazing things for which I really don't see any justification. Let me just slip in here a snarky comment stating that it is more expensive than the almost-identical regular table salt.

Apparently:

- It increases hydration. (Meaning, you can't do that with a glass of water?)

- It reduces blood pressure. Really? Better write a paper for the *Journal of American Cardiology*. They'll love to hear that. News to them.

- It is an air purifier. Does it have to be thrown out of an airplane?

I need a short rest before I can absorb more flimflam.

- It is a sleep inducer.

- It reduces aging.

- It—hold on there and sit down—increases libido.

Let me add my own with equal validation:

- It induces pigs to fly.

Bitter

What's my point?

Bitter receptors: here, there, and everywhere.

They are more about stayin' alive than anything else.

Bitter receptors do not work in a similar way to sweet receptors. They do not simply perceive bitter instead of sweet. They utilize very different mechanisms and pathways. There is only one "sweet" sensation and one receptor type for it (currently). If you think there are two, then I won't fight with you. Not so with bitterness. Much more research has been done on bitter receptors. There are more than two dozen types for different functions. You are likely to encounter bitter receptors almost anywhere in the body. Just to get you thinking along the correct lines, let me say, with a very, very broad brushstroke, that people with high densities of bitter receptors are less likely to suffer from chronic rhinosinusitis and some other diseases. The following is something that might not have crossed your mind recently: bitter receptors in the gut seem to play a role in alerting the body to fighting intestinal worms. But then again, intestinal worms seem to play a role in triggering an immune response to asthma. Nothing is ever simple, is it? The scope of this is well beyond this book or even my inquisitive brain. It does, though, allow me to yet again link taste receptors with homeostasis.

Bitter receptors and disease—let us discuss the bitter receptors that are in the head, mostly upward of the neck. The taste buds, which are very concentrated in the papillae nubbins on the tongue, are familiar to you because you can easily see and feel the nubbins by sticking your tongue out in the mirror (or even at someone you disrespect). Does everyone's tongue look roughly the same, or are some people's tongues rougher than others? The latter. Some people have a high density of nubbins. More nubbins mean you can detect tastes (especially bitter) easier and more intensely than those with a smoother tongue. Day to day, that does not matter much except that smooth-tongued folk are less picky with food than us annoying, rough-tongued folk. Let me state here that one is not necessarily better than the other. I am a pain in the butt to feed, but then again, when you do feed me, it does not require much to satisfy me. This is all a precursor to the section later that will be under "Supertasters," chapter 7.

The bitter receptors on your tongue do not only occur on the tongue. They occur less densely throughout the buccal cavity (mouth), into the throat, and up into the higher reaches of the nose. (As I write this, there is new research revealing that we have taste receptors throughout us, including the brain.) Now, we do not put lumps of food and drink up into our nostrils, so why did Mother Nature put bitter receptors up there? Mother Nature is not currently available for comment, but referring back to homeostasis (stayin' alive), it might give some clues. Let's ask Doc Rob (ENT), even though we don't have an appointment.

"It turns out that people with lots of bitter receptors are less likely to suffer chronic rhinosinusitis and other infections. How can that be? Bitter receptors detect bitter molecules. What produces bitter molecules? Bacteria infecting the nasal regions, for one. The bacteria produce bitter chemicals, but if these are detected quickly and efficiently, they can alert the immune system to get its act together and snap into action quickly and with full force. The region we are talking about is high up in the nose between the eyes. Homeostasis: everything trying to work together to produce optimum health and maximize the chances of survival."

Finally, don't confuse bitterness with astringency, as beer brewers notably are misguidedly inclined to do. Look at the astringency section in chapter 7 to learn why—especially if you are a drinks professional.

Sour

Evolution in action—we're not supposed to like sour. Vomit = sour. Sour has negative connotations. "He has such a sour personality—unpleasant and unfriendly." Compared to sweet, salt, and bitter, sour is relatively easy to explain. Before I forget, sour is modified (dimmed) by salt. This has reference especially to grapefruit, which I already explained. Sour is an expression of acidity, which means pH, which is a comparative scale using dilute hydrochloric acid as the reference point. Bitter uses a scale based on quinine or caffeine for reference. "Tart" is basically a food term for sour, likewise for "tangy." Here is a piece of advice: be careful how you use the word "tart" in the UK and its colonies. Yes, it may describe a pie or flan, but it can also refer to a lady who is, shall we say, loose with her morals and body. The village tart. Not the village sour.

Most people link sour with citric, and so they should. Citric fruits are citric because of citric acid. That wasn't too hard to understand, was it? Hereafter, confusion often creeps in between what is sour and what is bitter. Let us take the grapefruit as an example. The "meaty" (arguably inappropriate word to use in this case) part of a grapefruit is sour and sweet. The white parts, often called the pith, are bitter. In other words, grapefruit fruit is not bitter, but it is packaged by nature in bitter tissue. Citrus peel is bitter. As mentioned earlier, this bitter can be mediated by salt.

Referring to the opener of this section, sour has negative connotations, but we have learned in modern times to really enjoy the sour foods that, on an evolutionary level, we should eschew. As I write, sour beers are all the rage. These beers are heavy on lactic and other acids plus items which include *Pediococci*, which are—yes, you guessed it—foot bacteria. We enjoy this. Lactic acid enhances many foods in a way that acetic acid (vinegar) does not, although we have learned to enjoy vinegar with fish and French fries, or even chips. Why? I don't know. Habit perhaps. In Ethiopia, a food staple is their fermented teff grain flat bread called *injera*. What confuses flavor scientists is why young kids seem to enjoy extremely sour candies. Don't ask me; I do not know.

Umami

Personally, I do not think so. Not a valid taste. If we let that one slip by, then what is next? Electric? Carbonation? Fatty? Worst of all, metallic, which consumes and expansive section later on. The foregoing has all been posited but not universally accepted. As humans, does it matter much if we cannot instantly identify sweet, salt, sour, and bitter? Yes, if you like living. What about umami? For me, I cannot even identify it and don't know anyone who is able to, so it has no place in my life, yet I have lived for seventy-six years. How about other mammals? Cats can't give a rat's ... (questionable choice of phrasing there) about perceiving sweet. Why would they? They are obligate carnivores. Dogs, on the other hand, are omnivores. They'll eat whatever we eat. They don't care about over-salty foods, but sweet cookies are another matter.

I performed a Gerry's kitchen science test on some professional drinks writers. They were offered to each lick a spoon with some paste on it. No other information was proffered. It was commercial Trader Joe's Umami Paste, containing tomato, vinegar, lemon juice, grape must, anchovy, parmesan cheese, mushroom, and garlic. Their response was a surprise on many levels. None of them identified vinegar, despite it being considered an indicator of spoilage. Only lemon and mushroom were identified. No anchovy. None mentioned umami, and they laughed when I told them what they had just tasted.

MOUTH TRIGEMINAL PERCEPTIONS

What's my point?

Traditionally accepted tastes are just a part of what happens in the mouth.

Let's dump that unhelpful standard term "mouthfeel" and bring in "trigeminal response." Why? With mouthfeel, what does the mouth feel—anger, frustration, love, rejection? So, your gob-hole, as it was so classily called when I was a boy at Colchester Royal Grammar School, senses what goes into it. The nerves in the buccal cavity respond—notably, but not exclusively, the third branch of the trigeminal nerve. Doesn't that make more sense than "mouth sense" or "food texture"?

I am referring to the trigeminal nerve—as nerves go, a honking great python that snakes through many parts of the body, including the face and mouth. The "tri" reference has nothing to do with the root of "trivial." Just think of three. It covers much of the face. One branch around the eyes, one branch below the eyes but above the mouth, and one branch below the mouth (lower jaw), approximately. Don't get wound up about the details. If you really want to know, just look it up when you check out the cingulate gyrus mentioned earlier. Still don't get it? Well, perhaps it's time for a photo (Figure 45).

Figure 45.

The photo has my fingers representing branches of the trigeminal nerve that interest us when assessing taste. There is another branch that goes round the eye region, which comes into play when you cut onions and your eyes water (or get pepper-sprayed).

On a human, practical level, my trigeminal nerves tell me that I want to gag if tomato skin sticks to the roof of my mouth. I am not alone apparently. Let me back up a bit before I explain more about "trigeminal." To be honest, I chose this word because people remember it so easily, whereas they don't remember things like "olfactory epithelium." See how I keep repeating themes and messages? That's how I get my point across. Hit them over the head with it, Gerry—repeatedly.

By now, you will have assumed that I have pushed aside the old, simplistic "flavor = smell + taste" idea that is promulgated in just about every other source. If I say, "plastic + TV signals = CNN TV news," it would be true but somewhat unhelpful.

On an evolutionary level, what is the big deal with this powerful trigeminal response? Is it to help identify a wine like Château Lafite Rothschild Carruades Pauillac 1943? No, it is to assess the desirability of swallowing whatever it was that you just stuffed in your gob-hole—something nutritious, not toxic or irritating like a baby hedgehog. This process, of course, demonstrates Gate 3 of consumption: maintaining homeostasis.

The lips, cheeks, and tongue are loaded with sensory receptors. Speaking broadly, sensitive to what? Touch, temperature, pressure, texture, and pain, among others.

Look up "chemesthesis." Okay, I'll do it for you from Wikipedia.

> Chemesthesis is defined as the chemical sensibility of the skin and mucous membranes. Chemesthetic sensations arise when chemical compounds activate receptors associated with other senses that mediate pain, touch, and thermal perception. These chemical-induced reactions do not fit into the traditional sense categories of taste and smell.

Included in this mouthfeel group are the following:

- Astringency, which is understood by almost no food and drink professionals in my experience, despite what they tell you. Huge amounts more on this later in chapter 7, craftily hidden in the section called "Astringency."

- Temperature: receptors for hot and cold alerting you to too hot and too cold. Don't burn yourself because we need you to survive. Hot coffee and cold ice cream.

- Touch, including the feel of a utensil. Crispy/crunchy. Where would chips be without these?

- Burn, as in chili-pepper burn, which surprisingly is not dangerous even if overdone.

- Pressure: chewy, liquid(y), runny.

- Pain: chilies and fish/chicken bones.

- Carbonation: bubbles! By the way, you can get the feel of bubbles without bubbles. The miracle of carbonic acid. Animals don't like carbonation, so don't offer your dog soda water.

- Fattiness: very helpful in cheese, butter, and ice cream.

- Pungency and spiciness: biting, acrid, sharp, piercing, piquant, zesty, or burning. Question: What is the difference between a spicy burn, a temperature burn, and a chili burn? Answer: I have no clue. That's a little job for you to solve. I cannot be expected to do everything for you. In other words, anything that might be construed by the delicate among us as slightly offensive—mouth stuff, not smell stuff. I'm waiting. In the meantime, chew on hot chili. Mmm, capsaicin.

- Metallic: which is perceived by almost everyone, yet I reject it on the grounds that it cannot exist in biology, Your Honor. Refer ahead to the long, rambling chapter on metallic later in chapter 7. I think you'll find a lot of good, fun stuff in there.

So, all that foregoing is sensed by our trigeminal nerves. These nerves also allow us to chew and move the food bolus around our mouth using our tongue. That last statement has more going for it than you might think. We are the only mammals that can do that. Yes, I know that your dog can stick its tongue in and out, but it can't roll it and do disgusting things with it to amuse the children. Of course, without the tongue, our speech would come out sounding rather odd and prevent us from blowing raspberries.

To wind this up on a depressing note, have you ever heard of something called "burning tongue syndrome"? It's where you may experience a feeling that your tongue and other parts of your mouth have been scalded. Not reprimanded—scalded, as in boiling water. Your entire gob-hole is painful. Don't come to me for answers; my job is just to deliver the bad news.

"Time to lighten it up a bit, Gerry," I hear you say.

Should you be someone who likes diverse, pleasant trigeminal stimulation, then I have a gift for you: Szechuan peppercorns. Consume and discuss.

Thirst

How about a delicate tingling in your mouth? Like champagne, yes?

It will knock back your thirst, at least, and make your partner look even more attractive. If your partner is an athlete, then consider the following: sportspeople don't like to feel thirsty. There is a difference between feeling thirsty and needing liquid. To quell thirst, consume cold, carbonated liquids. This alleviates the sensation of thirst before the liquid is actually absorbed by the body. However, if you are more than thirsty—in fact, dehydrated—drink warm, still water. Now you really are confused.

Let's recap:

King Flavor =

- Smell
- Taste
- Emotion
- Homeostasis (no, you won't read that elsewhere)
- Gastropsychology
- Mouth trigeminal perceptions
- Psychosocial contributions

Remember, as mentioned earlier, "trigeminal" for me is a concept and embraces other nerves—notably the vagus nerve. Okay, you have been very patient on this dry subject, so I will reward you with a true story. In the operating room, I was with neurosurgeons who were going to implant a vagus nerve stimulator device. They were studying CT scans. Sidling up to them, I interjected, "Gentlemen, the vagus nerve has several branches. Collectively, they are known as Las Vagus."

Considering that it was original thinking, whereby I developed my "mouth trigeminal contributions" idea, it is my duty (and thrill) to explain it. No one else will, after all. This concept is rich in heavy-duty nerve/neurological content, but I will endeavor to get my point across without first insisting you complete first-year medical school or more likely have you turning pages until you get a nice picture to look at instead. Here are some instructions for you. Actually, consider them orders. Get away from thinking that a particular flavor triggers a particular receptor which goes to a particular part of the brain. It simply does not work that way. Luckily for you, it is easier than that, and anyway, ultimately, it is BobHo's job, not yours. Day to day, it does not even matter much.

Astringency

What's my point?

You haven't got a clue what it is really, have you?

Clicking your tongue on the roof of your mouth is not good enough.

Food and drinks experts are the only people who seem to be interested in astringency. They use the word and write about it. Why, when they can't even detect it even when it is the only ingredient in plain water? Okay, they can detect it in tea that has been steeped for seven days, as is normal in England and with certain new-crop red wines. As a national beer judge, presenter for the American Specialty Coffee Association, and educator to wine sommeliers, astringency pops its annoying head up for all of them. When I conduct a real blind tasting, none of them can identify astringency. This goes hand in hand with metallic.

In contrast to metallic taste, with astringency, we know what it is and what produces it. It is real, so there is no need for phantom experiences involving gnomes and fairies. People describe it mostly by facial expressions with tongues clicking upper palates and other distortions. Tea bags and grape skins are mentioned too. I have never met a drinks industry professional who can go further than that. The thing is that it is very easy to explain. If you insist on getting completely technical with it, then consider the following. (I'd skip it if I were you, because astringency is all based on a big, fat lie.)

Okay, I get it. What is astringency?

Now is the time to pour yourself a drink.

Astringency is a trigeminal sensation that involves the activation of G protein–coupled signaling by phenolic compounds. (But you knew that.)

Author information:

Schöbel N1, Radtke D2, Kyereme J2, Wollmann N3, Cichy A4, Obst K5, Kallweit K2, Kletke O6, Minovi A7, Dazert S7, Wetzel CH8, Vogt-Eisele A2, Gisselmann G2, Ley JP9, Bartoshuk LM10, Spehr J4, Hofmann T3, Hatt H2.

(The following is the kind of abstract which is mainly of interest to the author).

Abstract

Astringency is an everyday sensory experience best described as a dry mouthfeel typically elicited by phenol-rich alimentary products like tea and wine. The neural correlates and cellular mechanisms of astringency perception are still not well understood. We explored taste and astringency perception in human subjects to study the contribution of the taste as well as of the trigeminal sensory system to astringency perception. Subjects with either a lesion or lidocaine anesthesia of the chorda tympani taste nerve showed no impairment of astringency perception. Only anesthesia of both the lingual taste and trigeminal innervation by inferior alveolar nerve block led to a loss of astringency perception. In an in vitro model of trigeminal ganglion neurons of mice, we Cganglion neurons showed robust responses to 8 out of 19 monomeric phenolic astringent compounds and 8 polymeric red wine polyphenols in Ca^{+2} imaging experiments. The activating substances shared one or several galloyl moieties, whereas substances lacking the moiety did not or only weakly stimulate responses. The responses depended on Ca^{+2} influx and voltage-gated Ca^{+2}, but not on transient receptor potential channels. Responses to the phenolic compound epigallocatechin gallate as well as to a polymeric red wine polyphenol were inhibited by the G s inactivator suramin, the adenylate cyclase inhibitor SQ, and the cyclic nucleotide-gated channel inhibitor l-cis-diltiazem and displayed sensitivity to blockers of $Ca^{+2}Cl^-$ channels.

Don't you just love those adenylate cyclase inhibitors? Let's put it in plainer English, shall we? An astringent reacts with body tissues to shrink or constrict them. It coagulates proteins. So, gentlemen, after you shave your armpits and there's that irritated feeling and a little blood, and your guest in the motel with you screws up her or his face just looking at you, you need an astringent. Astringents occur naturally in some plants, as well as in alum, vinegar, rubbing alcohol, and a slew of other items. Famous sources include witch hazel, unripe banana skins, used tea bags (there you go putting unnecessary things in your mouth again), and tree tannins. This is where you can't help yourself from bursting into the Christmas song, "O Tannenbaum"— literally, "O fir tree." It is no coincidence that you can tan leather with a tannin. This brings us to oak tannins and of course wine and oak-barrel aging. Wines also get their tannins from grape skins, seeds, and stems. Lovely how I managed to link music, wine, tea, and expensive leather shoes, don't you think?

What Is That Tannic Taste?

Some red wines high in tannins are said to impart a dry, puckering sensation. Last time I checked, wine is extremely wet, so how come the dry sensation? Simply put, the wine is not dry; your tongue just thinks it is dry because of its constricted mucus membranes. It innocently lied to you. So how difficult was that?

Figure 47. "Dry...yet somehow, rather wet."

If astringency is a not a taste, what is it? I will tell you. It is a sensation identified by the trigeminal nerve. The trigeminal nerve basically handles sensory input from the side of the face and into the mouth. Here's my advice: don't chew on meat to find astringency. You won't find it.

Metallic: Just a Brain Phantom?

What's my point?

Is it just love and jealousy?

Here's what we know about "metallic taste."

1. There are two words.

2. It has more to do with linguistics than biology.

3. No one has ever tasted actual hard, shiny, elemental metal, which is what people think of when they hear "metallic": aluminum, gold, copper, iron, and silver.

Here, we must ask something that is not immediately obvious. What are metal and metallic? My questioning of diverse segments of society, including tribal Ethiopians, shows me that there are three broad groups of people, and here is what they think when they hear "metallic."

-   If you ask an astrophysicist, she will include oxygen, carbon, nitrogen, and sulfa.

-   If you ask a research sensory scientist, you are likely to have her talk about metallic compounds.

-   If you ask sensory science students, cooks, hobby wine tasters, and the remaining 99 percent of the population, they will think in terms of hard, shiny metals typified by coins, spoons, musical instrument mouthpieces and aluminum foil.

By the way, humor me for a moment and open a new box of foil and take a whiff. Smells of cardboard or nothing at all, does it not? Just so that we are clear, I am only talking about hard, shiny metals, because that is what comes to mind by just about everyone when I mention metallic. A famous PhD sensory scientist hit me with the heavy science by asking, "Haven't you ever sucked on a tarnished silver spoon?" No. Tarnished spoons are all he's got? Notice that he said that the silver was tarnished. That means it is not the element silver, it is one of thousands of possible silver compounds. It is not hard, shiny metal like aluminum foil. When I took the intellectual reasoning a stage further, I hit a brick wall with him. "If you can detect silver metal in a silver compound, does that mean you can detect the carbon in carbon dioxide?" Bam! Gotcha. No response, not even laughter. Here's a quick thought, did you ever hear a lady complain about the metallic smell of her new wedding ring?

Let me lay this on you. No one in the world has ever tasted or smelled elemental, hard, shiny metal. So, with all the words at our disposal, why do we select the word "metallic" when eating spinach (or whatever)? We also don't know what pandas' toenail clippings smell or taste like, but we don't say "pandas." This argument really annoys many people who should know better (than to mess with me).

Metallic Flavor or Taste

Most people will say that they have experienced a metallic taste. This includes my wife, who is amazing in every way including her choice of life partner. This can be from food, drink, intravenous drugs (especially cancer drugs and antibiotics), pregnancy, local anesthetic, taste in your mouth after a hard run, sucking on pennies, nuclear fallout, kiwi fruit, canned tomatoes—you get the idea. Ad infinitum, as the Romans apparently used to say. In fact, a host of things—almost anything. The most wonderful and obtuse for me was in a Netflix episode of the comedy series *Schitt's Creek*, wherein the love-hurt son said, "I can taste metal." All this gives me, as a writer, many problems, because we now have the dichotomy of metallic taste not existing yet people telling me firmly that it does. I am one of them, which makes it doubly troubling. A problem to the power of six?

Can it be that we experience something that we name metallic which has nothing whatsoever to do with metal? Examples of metallic-tasting items:
- Some fungi and plants (broadly)
- Kiwi fruit
- Spinach
- Dentists' injections
- Cancer drugs
- Anesthetic drugs
- Antibiotics
- Vigorous running
- Coins/blood (read on)
- Dementia/Parkinson's disease
- Pregnancy
- Coffee/beer

Let's stop there. Basically, anything that goes in the mouth or not. Anything.

Sorry, I am not finished with this, because there is another source of metallic that my audiences love to hear me talk about: post-orgasm metallic taste perceived by some women. That was guaranteed to get your attention. It seemed to me I should do a study on this. However, despite that I found plenty of lady volunteers, the study design hit a hurdle right there. No, I am not going to tell you what.

Blood Tastes Metallic

You get a cut on your finger, and then you reflexively suck the blood to stop it spilling on your crisp, white shirt. "Aha; that tastes metallic." I reflexively ask, "Why metallic?"

"Well, it's the hemoglobin."

"How do you know? Why is it not the hormones, enzymes, antibodies, proteins, white cells—you get the idea?"

"Because my mum told me it was the hemoglobin, and she worked in a butcher's shop." Let's look at this scientifically (sorry, all you sensory scientists).

The only part of blood I have heard used as a source of metal is hemoglobin. That's a third of blood. Don't ask me, "Wet or dry?" Sit up, all you flavor chemists; here's something for you.

Nerd chemically, hemoglobin looks like this:

$$C_{2952} H_{4664} O_{832} N_{812} Fe_4$$

Or graphically like this:

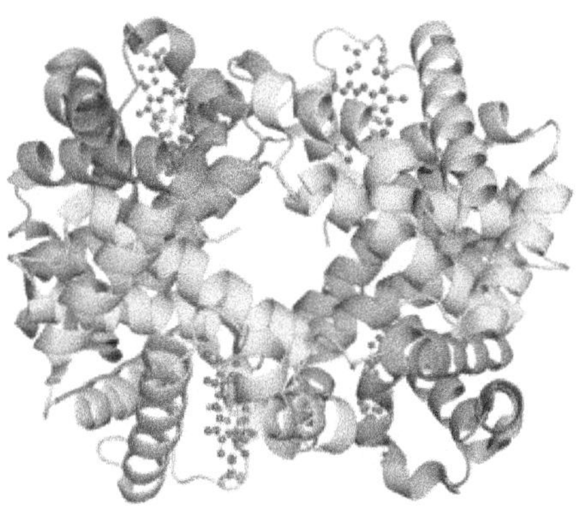

Figure 48. Spot the iron. Go on!

Oh boy, that Fe_4 sure stands out. This is a prime example of flimflam. You are welcome, Penn & Teller. Let me put it in a volume way. For years, I worked in the operating room with plenty of blood splashed around. It was like a shallow sea of blood sometimes. Not the Dead Sea, we hoped. Did anyone smell blood or metallic? Nonexistent. Today, I even asked a phlebotomist if she had ever smelled blood. After lots of hemming and hawing, she said, "No, and definitely not metallic. Next!"

Figure 49

Aluminum Foil Tastes Metallic

No, it does not. What is this obsession with chomping down on foil? "It's like when you bite on the foil of a baked potato." I would respond, "Really? Had you not noticed that foil is a different color from potato?" If there were only occasional respondents who came up with the blood/foil thing, we could ignore it as aberrant human behavior, but I hear it all the time. Something is up. What is it? I don't really know as I write this, but here's another common one for you to ponder...

Sucking On Metal

What's my point?

Just about everyone has an opinion on what "metallic taste" is.

The two most common responses, in my experience, are connected with drinking blood and eating aluminum foil. What's wrong with you people? Let's just dissect the two scenarios.

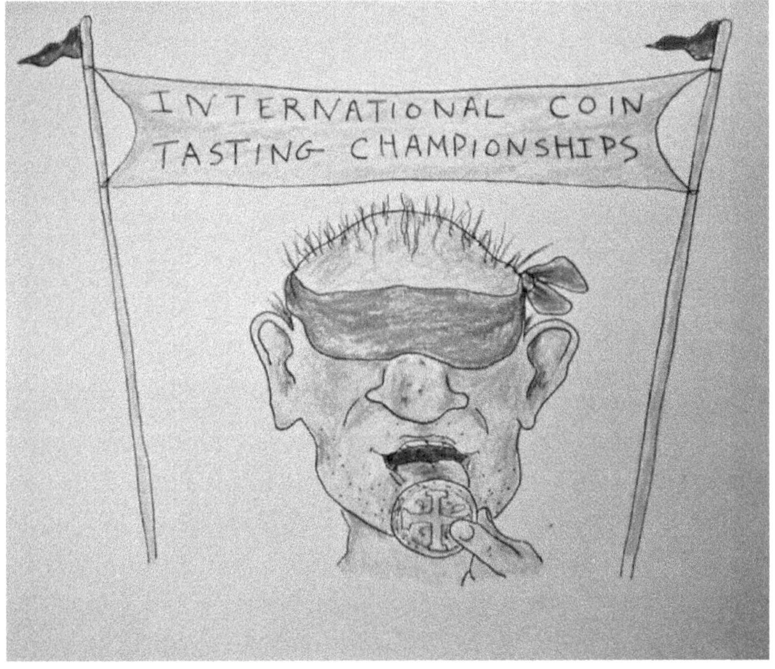

Figure 50. *"It was Carlos who won, correctly identifying the 1537 Spanish doubloon."*

"It's like when you suck on a penny." Here's some news for you: you are not three years old any longer. You don't have to suck on everything that's not tied down to your stroller or Harley. I also have some bad news for you. Some bored scientists took you seriously. Cutting to the chase, they found that it was skin oils reacting with the metal that may give coins a smell or taste to those who can't resist sucking and smelling them. Fine, don't believe me? YouTube has great videos on it. Despite this, some people still say they can taste coins.

If you were thinking rationally, you'd ask what the big books on taste say on the subject. If you want a speedy exercise, then look up any authoritative tome. That didn't take long, did it? Don't be upset. It's getting time to do some deep thinking, because at the moment, we are wallowing away in uncertainties. Let's apply some logical, rational thinking to this subject. When in doubt, bring in the deductive reasoning (again).

Metallic Flavor or Taste Using Deductive Reasoning

Flavor = smell + taste.

But metals are not volatile, so have no smell, therefore can't have a flavor.

The only way we could detect metallic would then be by taste.

Taste = sweet, salt, bitter, sour, but not metallic. We have no receptors for metal.

Therefore, metallic taste can only exist as a phantom concocted by our brains. Unless there is something else at play. To explain further, one must accept that metals (apart from mercury) are not volatile at room temperature, so they cannot exude a smell detectible by humans. Charles Spence's *Gastrophysics* posits that there may be other tastes as yet undiscovered. He may even be staying awake at night hoping to witness a flying pig. Even if this is true, how does a medical drug in our bloodstream, pregnancy in general, or a long run convert into a taste in our mouth? I wrote to him regarding this, but he refused to discuss it. With luck, he never hears from me, Gerry Nicholls (truth-teller), ever again.

Roman Urns and Hair Dye

What metals do you think would give a metallic taste in drinking water? Iron? Lead? The Flint, Michigan, water crisis, where river water was eroding people's water pipes, produced colored, bad-tasting water. With their skanky, iron- and lead-infected water, the populous complained of metallic taste, right? Wrong. Let's make it even easier. In Roman times, they used lots of lead for directing water (plastic of the past) and storing wine and grape juice. If you store wine or grape juice in lead vessels, there is a known chemical reaction. Grape juice boiled down in lead vessels produces lead acetate $Pb(C_2H_3O)_2$, which you might expect to taste metallic. Wrong. It tastes sweet. They even had a name for it: sweet water. Lead compounds are bad for your brain and head, yet they are the basis for some hair dyes. Lead acetate = sugar of lead = a lead salt (the chemical, not the taste) = a sweet taste, which is unusual, because poisons are usually bitter. There goes that evolution again.

How far did ancient people go to enhance the flavor of their food and drinks? Would they consume toxic substances if it made items a little more appetizing? Well, the Romans did, by adding a sweet version of lead to their wine and, later, to their food. Some scholars say that widespread lead poisoning contributed to the fall of the powerful Roman empire. You think we are smarter these days? Think again. It took until the 1980s to phase out lead wine caps. So that's it then. Problem solved. Metallic taste does not and cannot exist.

At the dentist's last week, the local anesthetic tasted metallic to me. What a nuisance, because now I must accost my dentist friend and ask for a local anesthetic sample to play with (sans syringe), just to suck on. It tastes the same ingested or injected. "That's because just before injection, we squirt a little out before the needle enters the gum," said my all-knowing dentist. Ask a hundred different people what "metallic taste" is, and you will get a hundred different impressions. In desperation, I contacted my Ethiopian translator friend in Addis Ababa. They know about the term but can't go further than that. Even people who have no interest in food and drink flavor may complain of a metallic taste from medications, commonly antibiotics. Here's the thing though: the medication taste comes after it has been digested/metabolized, not perceived as it is swallowed.

2017 SmellTaste Conference

At the 2017 SmellTaste conference at the University of Florida, I took the floor and proffered that metallic taste was simply a phantom created in the brain. I asked for comment and got nothing but the sucking sound of dropped jaws from the brightest and the best. Could it be that part of the problem is that we are using the wrong word with "metallic"? Why not "pandas' toenail clippings"?

As I write, I honestly do not know what metallic taste is, but I feel strongly that we are using the wrong word to describe what we consider to be metallic. One thing that it is not is metallic. Dentists spend their lives in people's mouths, so I went to them again for answers—the dentists, not the mouths. A dentist and a dental hygienist friend agreed to my questioning them: After receiving metal fillings, how many patients perceive a metallic taste? The answer shocked me: never. Maryanne said, "They do complain of a nasty, bitter taste though." Dr. Duchan agreed: "After injecting local anesthetic and spilling a little out of the syringe before entering the gum, they say it tastes bitter and nasty." A light went on in my brain. Is the metallic descriptor in fact a form of bitterness? At least we would have a physiological explanation for the taste. It is known that bitter taste receptors are complex and ubiquitous throughout our bodies. Reread the section in this book on taste and testicles in chapter 5.

Before I get overly happy with my brilliance, this still does not explain the blood, drug, pregnancy, and mixed martial arts scenarios whereby nothing passes through the mouth. Sit down. Many women after sexual orgasm experience a metallic taste. Please refrain from telling me that condoms are made of copper.

Love and Jealousy

I can see that you are getting tired, so how about a bedtime story? No, not you going to sleep. I am talking about the story of the other night of me in bed with my wife. She was reading *New Scientist* journal containing an article regarding how humans develop language. "This might help you," she said, not realizing that those four words would send me on an international quest. My destination became Dr. Freddy Jackson Brown (I don't think he's the famous singer) at Warwick University England. He is a linguist and loved my cartoon and was kind and patient enough to listen to my plight. After some back and forth, the question arises, if metallic taste does not come from actual chemicals and nerve pathways, perhaps it is a word that some of our brains use to capture the concept enough to think it is real—like love and jealousy.

Metallic taste is a fantasy, an illusion, a phantom which we perceive as real. But why that word? A metallurgist friend of mine expressed it this way: "Metallic taste can't exist. However, we say things like 'It is raining cats and dogs' with full knowledge that there are no wet mammals plummeting down from the sky."

If we look at a sheet of cold, hard metal, we don't think in terms of cuddly sweetness. We say that something smells sweet. Nothing smells sweet. Sweet is a taste. We may smell something that reminds us of something that we know was sweet by experience—like a ripe banana. But metal is cold and hard and probably does not have a pleasant, heartwarming taste, so if we smell or taste something offensive that we can't really identify, the closest we may get is to call it "metallic." If it is nasty in a putrid or fecal way (it tastes like s**t), then we name it as such, because historically, fecal, and putrid molecules existentially have traveled into our noses. Yes, farts are real molecules—molecules, not just words. Get over it, and giggle with your offspring. Metal molecules do not enter our noses, because metals are not volatile at room temperature. All we know of metallic is what we think we know. In science, we like reproducible exercises. To confirm, don't ask the millions of brass instrument players if they can taste metallic when they play—unless they are playing the band Metallica's music.

Martial Arts Pain

As I hope we have established, we do not even have to ingest anything to experience metallic taste. Take the case of mead-maker Will Covill, who confided in me, "Back in 2014, I was a participant in the Zandri's Martial Arts MMA program in Brookfield, Connecticut (now Hammerhead Martial Arts). One night, we were working on grappling, and my sparring partner managed to wrap both of his legs around my upper chest/rib cage. I was unable to dislodge his legs, and he managed to squeeze my ribs inward.

It was a seriously curious sensation, as there was absolutely no pain, and yet I felt a moment of panic, an objective understanding that I was close to serious injury. I immediately tapped out, and as he released the hold and my rib cage expanded back, there were two very distinct feelings. One was a flood of pain in my bones. The next was a distinct, strong taste of copper in my mouth, literally like pennies and blood. It stuck in my mind because of its weirdness, and to this day, I can remember the sensation of my bones flexing and the strong, bizarre taste."

There is an evolutionary advantage to being able to detect tastes. Sweet indicates high energy food; bitter warns, "Beware. It may be unripe or toxic"; sour warns, "Beware, because it may be infected or spoiled"; salt is needed by all. Hold on though. Ten thousand years ago, all the metal was locked underground, and early man had no access to it. Not trusting myself, I contacted Peter Unger, professor of anthropology at the University of Arkansas and an expert on primate feeding ecology. He could surely help me. I asked him why humans would want to be able to detect metal in their diet. He replied confidently with a raucous laugh and the statement, "I have no idea, and I don't know who might. Great question though, which you should follow up on." They call this a dead end whilst kicking the can down the road.

Selling Off-Flavors

People who devote their lives to identifying tastes, including wine sommeliers, coffee cuppers, beer brewers, honey producers, and all forms of chefs, consider that a metallic taste is a fault. They are so confident in this that they train themselves to detect metallic faults. They even have little vials of liquids to train on, so let's examine this avenue.

The coffee and beer self-professed experts use compounds to demonstrate metallic but do not utilize the same compounds (typically ferrous sulfate and potassium chloride), not even close. One is inclined to ask why they have not chosen the same compound or something obviously metallic to make the demonstration, like nails, copper sheet, coins, aluminum foil, or even daggers.

The Siebel Institute of Technology in Chicago trains beer brewers and yeast-expert wannabes. They are very respected by the industry, and rightly so. They have a great sales plan, whereby they will sell you a kit to train yourself to identify some tastes that don't exist, including a Beer Advanced Off-Flavor Kit for $176 in 2023 money. To your heart's content, you can spend lazy weekends with your honey smelling butyric acid (putrid baby vomit), isovaleric acid (cheesy, sweaty socks), indole (farm, barnyard), mercaptan (sewer-like, drains), and on a delicious note, hydrogen sulfide, which is the ever-resistible rotten eggs.

Metallic is also discernible thanks to the joy of chewing on anemic pregnant women's iron tablets, which are simple ferrous sulfate. I have blind-tested dozens of brewers with this, and not one could detect it. One would be forgiven for assuming that the metal taste is from the ferrous part of the compound. But why not the sulfate part? "Never thought of that," an expert told me. The thing that confuses me is that we are supposed to be able to taste the ferrous in ferrous sulfate but not the ferrous in sucking on a piece of good, old 100 percent iron rod. How do I know? I sucked my way round the metal aisle of Home Depot with my wife as lookout in case they took me away. On second thought, it might have been worth it just to see the judge's face. Let's look at another off-flavor kit example, this time from the Specialist Coffee Association of America, for whom I am a presenter. Take into consideration that green coffee is the most-traded commodity after oil.

Gullible Beer and Coffee Judges?

The American National Beer Judge Certification Program certifies beer judges who attend competitions. Nobody forces them to volunteer—they do it for fun, would you believe? The official judge form now includes directions on metallic and astringency. The 2016 World Coffee Research Sensory Lexicon (unabridged definition and references) tells us that metallic is an aromatic (how can that be if it is not volatile?) and mouthfeel associated with tin cans and aluminum foil. Hold on there; haven't we just said that metals are not aromatic, and cans are nowadays made of steel or aluminum lined with plastic? Open a new roll of aluminum foil and take a whiff. Nothing. Let us proceed with their method of duplicating metal taste: 0.1 percent potassium chloride solution. That's right—road salt. By the way, no one has ever tasted pure potassium without destroying the inside of their mouth. I tried to buy a couple of grams of potassium chloride from my pharmacist, but she turned me down without a doctor's prescription in case I overdosed. Yet I can buy forty pounds of road salt from my hardware store without a prescription. I wonder who thought that rule up? I made up the solution and it tasted like...dilute road salt.

The Beer Judge Certification Program (BJCP, which, incidentally, was started by my friend Pat Baker, who I think may have been an industrial spy—that's another story) offers fascinating directives. BJCP's Ted Hausotter offers this beauty: "Metallic is commonly described as inky." Well, whoop-de-do. Is it really? How many of you out there drink ink often enough to recognize it? "Mary, would you like a beer or a nice pint of ink?" That would be a quick way to ruin a relationship. Hausotter also offers, "Copper increases the astringency of beer. May taste like copper, but usually, astringency and harshness in character are the main attributes." Forgive me for saying so, but it seems that Ted does not know what astringency is. That is not really a shock when you read my foregoing chapter on astringency.

Ted tells us how to doctor a beer (even if it is not sick) with the following instructions: "Open a cap on a chilled beer." (Notice that he does not say what happens if it is not chilled.) "Put an iron or copper pipe into the beer for a set period of time—forty-five seconds." I tried this with a sheet of new, shiny copper—forty-five seconds, nothing, and four days, nothing, so I took it to my taste panel. Nothing. In an act of desperation, I tested the chief scientist of a corporation that makes gas chromatographs who is also a beer judge. He detected nothing remotely metallic from ferrous sulfate or potassium chloride, whereas I detected a strong smell of flimflam. On the Beer Judge Certification score sheet, there are sections on astringency and metallic. I offered to help them (free of charge) to devise a form that was scientific, factual, and helpful to the judges and brewers. No thank you, Gerry. We love living in the past in our flimflam world. Don't rock the boat.

Sweet, Salt, Sour, and Bitter Tastes

As you know, I am at pains not to rehash other people's work, and loads has been written on the tastes, so feel free to wallow in a big bath of sweet, salt, sour, and bitter far and wide elsewhere (with a side dish of umami, if you can locate it). It will quickly be revealed to you that authors are people repeating other people, repeating other people, repeating other people. Shall I repeat that? Let's get to something important and fundamental though: Why are these tastes so very, very important to us on an evolutionary level?

The key word is yet again *homeostasis*. Let me put it this way: our body likes to remain on an even keel. What do I mean by that? It likes to be within a degree of thirty-eight degrees Celsius. Notice that I stayed away from the archaic, illogical Fahrenheit scale. We operate more efficiently at the correct temperature for hormones and enzymes within us. Like Goldilocks: not too hot, not too cold. Just right. Collapsing immobile and having brain strokes are not considered to be good. Eating an amount of food that is nutritious and provides sufficient energy and protein is generally regarded as an excellent thing. That is why everyone is the correct weight. Back up there, Gerry; we all know that is not true. Humans are emotional animals. Emotion trumps common sense and self-preservation. I feel another book coming on. It is not hard for us to poison ourselves and become diseased by what we consume. Whether you like it or not, as you raise your glass of wine, alcohol is a slow-acting poison. Bad for homeostasis. The danger is in the dose size. The following is an extreme example of what homeostasis is NOT.

Go to the seaside and drink gallons of sea water (salt) then look around for some rotting squid or fish (sour). Be careful when consuming poisonous jellyfish (bitter) and buy an ice cream afterward (sweet).

Taste Receptors

There are specialized taste receptors in the mouth, buccal cavity, and lungs (yes, lungs). There is not, however, one receptor for sweet, one for bitter, one for sour, one for salt, and so on. Detecting bitter works in a totally different way from detecting salt or sweet. Go to a mirror and look at your face. Terrific specimen, eh? No wonder others find you irresistible. Destroy the sensual moment by sticking your tongue out. This does not require you to make a sound. Unless you are fifteen. I hope you are. You are the future—unless you don't consider homeostasis when thinking or doing anything. Back to the tongue. Call your dog over. Stick out your tongue and roll it. Watch your dog becoming consumed by jealousy because she can't do that.

Don't Trust All Dentists

Still appearing in literature and on dentists' websites are representations of the tongue map, show-ing where on the tongue we perceive each taste. It has all been repeatedly disproven and is simply uneducated nonsense. Other books spend a page on this map subject—even the ones denouncing it. Not me; we have better things to do. Where was I? Oh yes—taste receptors/buds. There are high concentrations on the little nubbins (easily understood, nontechnical term for papillae), bumps which hold the taste buds that make the tongue feel rough. These receptor sites extend from the mouth up into the nose, the rest of the mouth cavity, and all stations south, east, north, and west. Taste receptors are all over the place in our bodies. Let's entertain the fifteen-year-olds—and tes-ticles. In my talks, I have a section titled, "The Four Ts: Taste, Tongue, Texture, and Testicles." Full disclosure: adults laugh too. This is a fascinating subject, and I encourage you to look deeper into it. You can start with somewhat of a mouthful. Google, "Functional characterization of bitter-taste receptors expressed in mammalian testis." Don't read it. It is enough to know that I am not pulling your leg. There is increasingly common (and stronger) research supporting this and broadening it to all parts of the body, including the brain. Okay, you just cannot resist, so let me explain it in just a few words. Some research scientists were genetically altering mice (without their permission or knowledge) and tried to breed the line on.

Unknowingly, they had removed bitter receptors in their testicles (the mice, not the scientists). This produced sperm that was not up to snuff. So why would Mother Nature want to have control of sperm fertility? Come on, I just mentioned it a few paragraphs ago. That's it—homeostasis. Functional sperm good, nonfunctional sperm not good for the long-term prospects of the species.

In the scientific world, there is an ongoing quest to identify more and more tastes. Hold your horses, cowboy; I haven't finished with umami. Fat? Not just a textural thing. Oleogustus is the fancy name. Then there are the ever-popular kokumi, electricity, bubbles, calcium, water (!!!), soapiness, lysine, alkaline, hydroxide...oh please, let's just stop here. In my humble (yeah, right) view, I don't care how many or few tastes there are. You can decide for yourself. Make it your weekend project. I won't try to sway you except for my nemesis, umami. Give up boxing and knock yourself out with something less useful or healthy.

Complex Flavors: 1 + 1 + 1 = 1

What's my point?

Is complexity better?

If you want to look your most attractive, hang around ugly people.

Let's face it; sometimes the brain is hopeless. It can remember the lyrics to so many songs. It can remember many, many, many smells, and flavors, but if you go all sneaky on your brain and mix smells and flavors, it gets horribly confused and does not like it. I hear you say, "The average meal is a mix of many smells and flavors." Exactly. If I put a plate of food on the table with meat, dairy, vegetables, and sauce, then work up a mixed forkful and give it to you blindfolded, can you name every item on the fork even when they are all very familiar to you? Liar; no, you can't. Don't just sit there with your distrusting demeanor; try it on your partner but be prepared to sleep in different bedrooms.

Let me say that I have no proof a mixture is incapable of making your meal more enjoyable. I am just stating that you cannot identify the mixture.

For lunch, I had a toasted cheese and tomato sandwich. Last minute, I asked my long-suffering wife to add some fresh basil from the plant outside our front door. The leaves were as fresh and full-flavored as one could ever desire, but once mixed with the sandwich, the basil flavor ran off with its tail between its legs. It was not detectable. I admired it and desired it. The sandwich refused to understand the saying, "Hang around ugly people if you want to look your best." If it was put on a dry cracker, perhaps with some mozzarella cheese, it would have sung out. As it was, the cheese and toasty Maillard reaction on the bread overshadowed it. I felt sorry for it.

The brain is in effect saying, "Why do you try to overload me with sensory information? I am just a simple organ. I am not a jazz organist like Jimmy Smith, who can handle all that complex improvisation stuff with feet and hands. I need simple, repeatable rules. Don't try to confuse me with lots of input when I really don't care. Just give me one or two things at a time and let me enjoy them to their fullest. Mixtures just confuse me (in a sort of pleasant way, I admit) when it comes to identifying what I am consuming. I can enjoy and correctly identify tastes individually, but mix them, and you have screwed me. I know orange. I know vinegar. But mix them, and I am lost."

Let me give you a specific example to elucidate what I am saying. Hard cider is just apples plus the products of yeast fermentation. Hard cider is apples, apples, apples. Yet the marketing geniuses have decided that great apple hard cider is not good enough. They want to make it better, so they add stuff into a near-perfect and very balanced drink. Generally, they choose sour fruits like raspberry and cranberry. Let me tell you a secret: the brain does not like this, because more is not necessarily better. I was at a beer festival working (yeah, right) and took a sample of apple cider with raspberry to blind test all the brewer experts. I also did this with a simple sour beer with one fruit added (cherry). You can't wait for the result, can you? Not one professional brewer identified the lone fruit in the simple sour beer. When it came to the cider with raspberry, they either identified apple or raspberry, not both. What? The drink only had two things and Mr. Brain got confused after one? Correct. The brain is in effect saying to you, "What do you want, apple or raspberry? Make up your mind; you can't have both."

As you keep adding more and more ingredients/flavors, you get a more and more complex mishmash of things you can't identify. I would ask, "Why add expensive ingredients if they can't be detected?" When confronted, the chefs or brewers generally excuse themselves by saying that "they add complexity." I dispute that unless you are talking about visual complexity.

Even considering strongly trigeminally perceived items, is it more satisfying to have greater complexity? Does an oyster require more? Is a crispy, crunchy, fatty, spicy, slimy apple concoction more satisfying than a standalone apple? The visual aspect (gastropsychology) invariably injects a substantial role when we consume. How many of you eat and drink without looking first at the dish or drink? You see the item and from your memory make decisions as to what to expect. The brain immediately makes decisions for you based on emotion and memory. There are even entertaining examples of chefs being given what they are told are apples to taste, which are in fact entirely different—they just look like apples. The visual part of the brain wins out, and voilà, apple it is. Now before you get the impression that I am a big shot who would never fall for something like that, let me reluctantly—even ashamedly—reveal something to you. At a pub in England, which had a long row of beer handles, I tried a few halves of bitter. Purely intellectually, you understand. Bitter is a style of beer related to pale ales. Eventually, I asked to try the one at the far end. "What would this bitter taste like?" I wondered to my gullible self. It certainly was not as good as the others. It was a pleasant drink in its own way but not a bitter that I would choose to drink. I was drinking apple cider and did not realize it. I was expecting a bitter, and nothing would sway my brain away from that concept.

If you want to look your most attractive, hang around ugly people. What did I mean by that? If you want to appear the most attractive in a bunch of people, don't hang around film stars. Hang around ugly people. You will shine compared to them. If you want a food or drink item to knock your socks off, consume something bland first. Here is a real example: a British professional brewer told me he was visiting a pub which was not his local where he drinks the beer he brews. He had an IPA, which is much more bitter (not bitterer) and assertive than the regular pale ale he brews and drinks. After the IPA, he returned to his local and ordered his usual pale ale. It tasted "different" to him, much less assertive than he was used to. He calibrated his perception to the previous very bitter beer.

Let me put it another way. When we have a full meal, why don't we start with dessert and finish with a piece of dry bread? My view is that because we calibrate to whatever we have had previously, we would be progressing from assertive to blander. The meal would become less and less interesting. As always, there are exceptions. In this instance, Maria informed me that her old Chinese neighbor always started a meal with the dessert. That's humans for you—unpredictable.

Is More, Better?

What's my point?

Why is Gerry's limit (1 + 1 + 1 + 1 = 1 or less) less than the Laing limit?

Do you like to scream? This is for you. Instead of spending thousands of dollars, they could have just asked me, and I'd have told them. Humans are hopeless at correctly identifying items in a mixture. Usually, one only. Often wrong. Occasionally two. By the way, this study, in my humble opinion, is an example of terrible methodology. It would be very easy for me to trash it, but I can't find any humor in it.

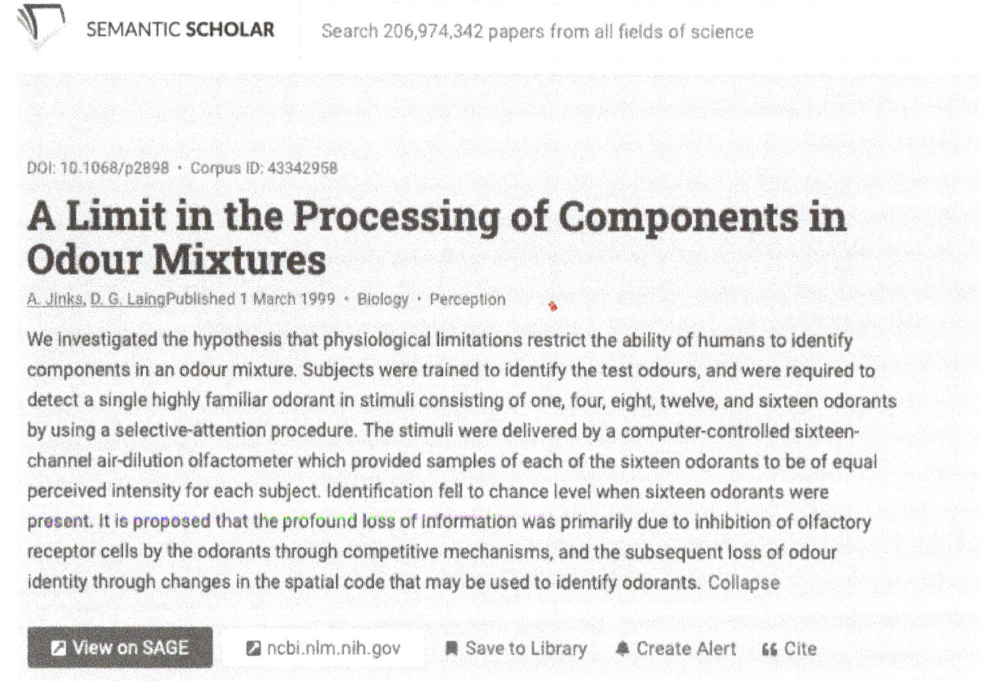

Figure 52

Are we kidding ourselves that more than three ingredients adds "complexity"—whatever that is? By adding more, are we dumbing down the essentials of gin, not complementing them? Are a hundred ingredients better than ten? Are ten ingredients better than two? Do gin formula developers even check? Not to my knowledge, because I asked some—see chapter 7.

Real World of Gin

My distilled-spirits writer friend came back from a trade show, and his comment to me was, "Gerry, you would have gone nuts seeing the ingredients they are combining in gin. And so many of them." My concerns with gin are that although we have limited ability to identify items in a mixture such as gin, we have more latitude when considering only trigeminal perceptions. Allow me to tackle this concept using breakfast cereal as an example. Your breakfast cereal dish may be perceived trigeminally as crispy, crunchy, cold, and, with yogurt, creamy. These have nothing to do with taste and smell. Trigeminally, gin is much simpler. In gin, we have alcohol, which is bitter, and a slew of spice and fruit possibilities, which embrace trigeminal, taste, and smell perceptions. With all that is already going on in gin, why is it a good idea to perhaps add orange and lemon? At this point, I come back to Gerry's Kitchen Science Experiments, which clearly showed that almost no one knows straight lemon from lime (and sometimes grapefruit), let alone combining these with all the other assertive gin ingredients often named "botanicals." Botanicals is a fancy way of saying "not meat." The other aromatics in citric fruits have nothing to do with the sour. They just separate orange from lemon.

Surely 13,475 ingredients would be close to the ultimate. If so, people must hate water, wouldn't you think? Does the gin distiller who proudly makes his balanced gin know that perhaps most consumers will immediately adulterate it with extra lemon and tonic? What an insult. Are all the ingredients beyond juniper and lemon just fluff to be able to say, "Look at our gin and all its amazing ingredients that other gins don't have. More is better!" If a gin distiller has a gin with ten ingredients, does he/she conduct blind tests to see whether if any one of them is removed it can be perceived? I nervously await your response.

Food and Drink Pairing

What's my point?

Everything goes with everything.

There exists now a food movement which experimentally combines various food items, dishes, and combinations of food and drink. Famously, wine and cheese pairing events come to mind. Because humans consume hedonically, we have a penchant for making everything as complicated as possible. Is it possible for anyone to say, "This or that pairs well (or badly) with this or that?" Does anyone know for sure? To be honest, I have dreaded writing on this subject, because personally, I simply do not understand food and drink pairing. As my drinks writer friend is wont to say, "It's no good asking Gerry; he doesn't drink with his meals." I drink when I drink, and I eat when I eat, and never the twain shall meet. It makes me odd and unusual; I know. For me, no drink pairs with food. I don't drink anything (apart from stimulating conversation) during eating.

In John McQuaid's book *Tasty*, he mentions a Belgian company called Foodpairing. By the way, I tried contacting John, to no avail. I tried. This led me to an article that was published in the esteemed science journal *Nature*, "Flavor network and the principles of food pairing," by Ahn et Bagrow et Barabási Scientific reports 1, Article number: 196 (2011).

What does the paper conclude?

I have no idea, but the graphics are really very pretty. Here they are for you. They elicited some comments from food/drinks writer friends, which include, "Someone has too much time on their hands," "Wow, someone wasted a sh*tload of time on that diagram," "Now I really don't understand," and finally, this gem from a British brewer, "What a complete load of bollocks." Judge for yourself. Here are the bollocks:

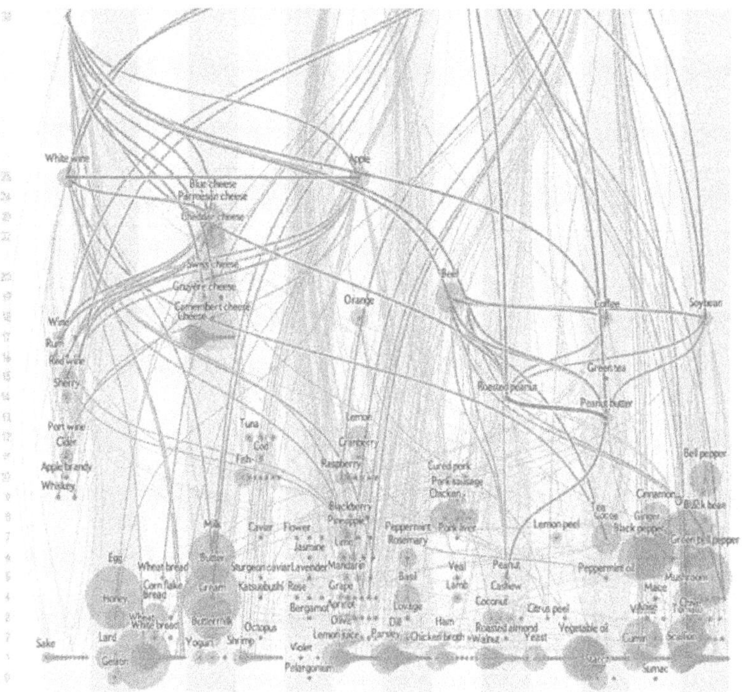

Figure 53

There exists a synesthete (see later in the chapter "Synesthesia") who I am sure would laugh at this. This synesthete's idea of a balanced meal includes "baked chicken with vanilla ice cream (on the same plate) plus orange juice. Pumpkin pie adds purple to this delicious combination. If orange color is required, then add raspberries. Raw spinach will add more purple." Yes, we surely all live in our own flavor world.

The authors do admit that what pairs well in the West is not what pairs well in the East. But what about all immigrants? Does this mean that pairing is not something that can be agreed upon? Can you explain this to me: if you eat some curry and follow it immediately with ice cream, you are essentially pairing curry with ice cream. If this is so, why not have them on the same plate together?

Wine Pairing

Perhaps because there is potentially so much money to be made and because it is so easy to execute (not in the slaughter sense), wine pairing is big business. Typically, "experts" like Wine Folly's Madeline Puckette have all kinds of information available on the internet and elsewhere. They proffer expert advice on how pairing works—what pairs with what and how to taste wine. I can't prove her wrong, but getting to basics, she does not even seem to understand basics: the difference between bitterness and astringency. She does not seem to understand that alcohol is bitter. The basics have eluded her. Not uncommon. She did not respond to a communication from me. She joins the queue (or is it a stadium?) now.

Fusion Cuisine

What's my point?

Is it just confusing (and fusing) flimflam with ignorance?

Figure 54

Is fusion cuisine a form of food pairing? What are we talking about here? Fusion cuisine refers to varying cultural and geographic dishes and ingredients being "paired." For example, this occurs when geographically separated Southeast Asian islands share dishes. California uses the influences of Italian, French, and Mexican traditions to fuse dishes. Even the apparently foolproof pizza has been the subject of fusion. Pizza with Japanese sushi toppings or Irish cheese are examples. Okay, put some Mexican beans on there too. One could say that fusion cuisine utilizes varying culinary traditions and ingredients. Anything goes—or is there someone writing the rules on pairing? Probably not, because apparently, Asian people disagree with others' ideas of what pairs with what. Here is an odd way of approaching pairing: an analysis of recipes showed that jasmine tea and thirteen other common ingredients appeared infrequently in recipes. Eggs, however, appeared in a third of all recipes—tens of thousands. So, if in doubt, chuck an egg in. Let's keep things simple and logical by just following whatever the following graphics are trying to explain. While you are doing that, I shall have a cup of tea with nothing in it. It's going to take you a while.

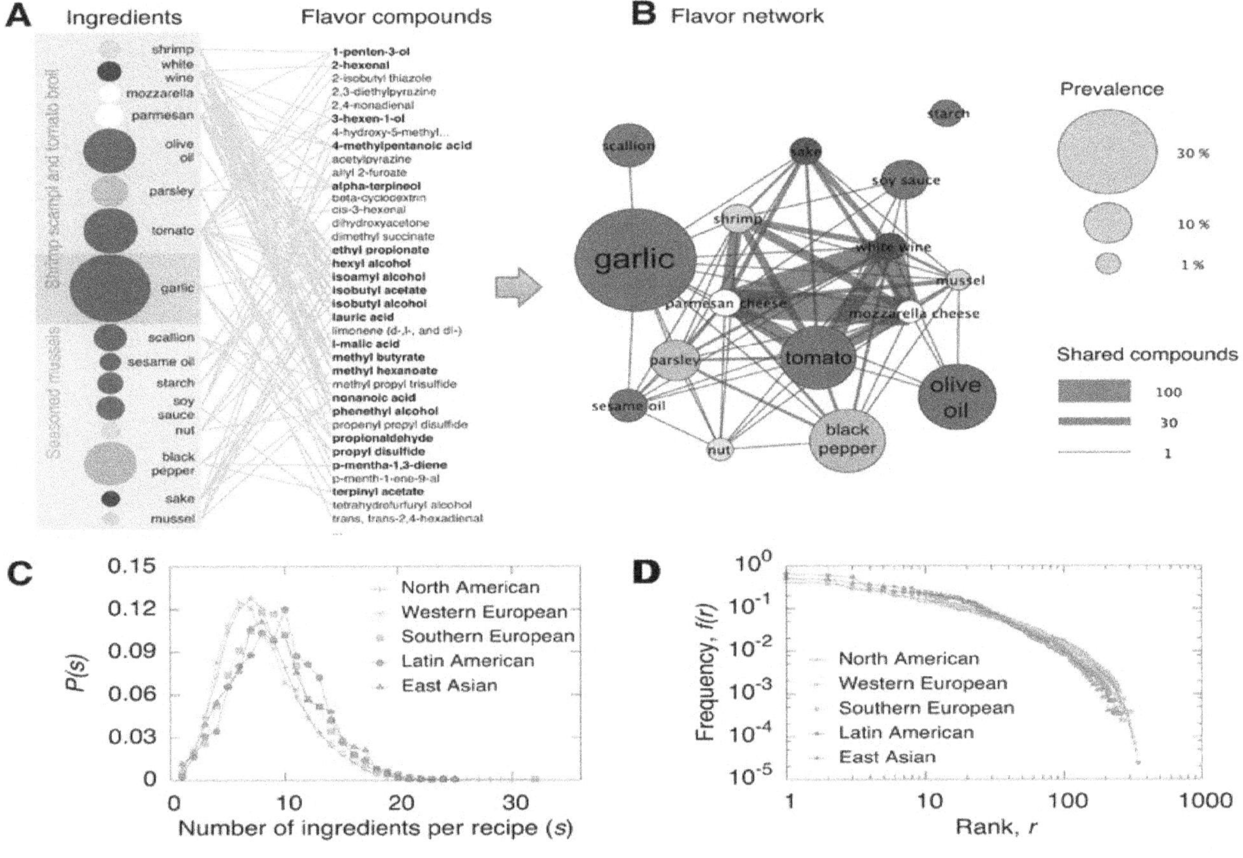

Figure 55. I don't understand either.

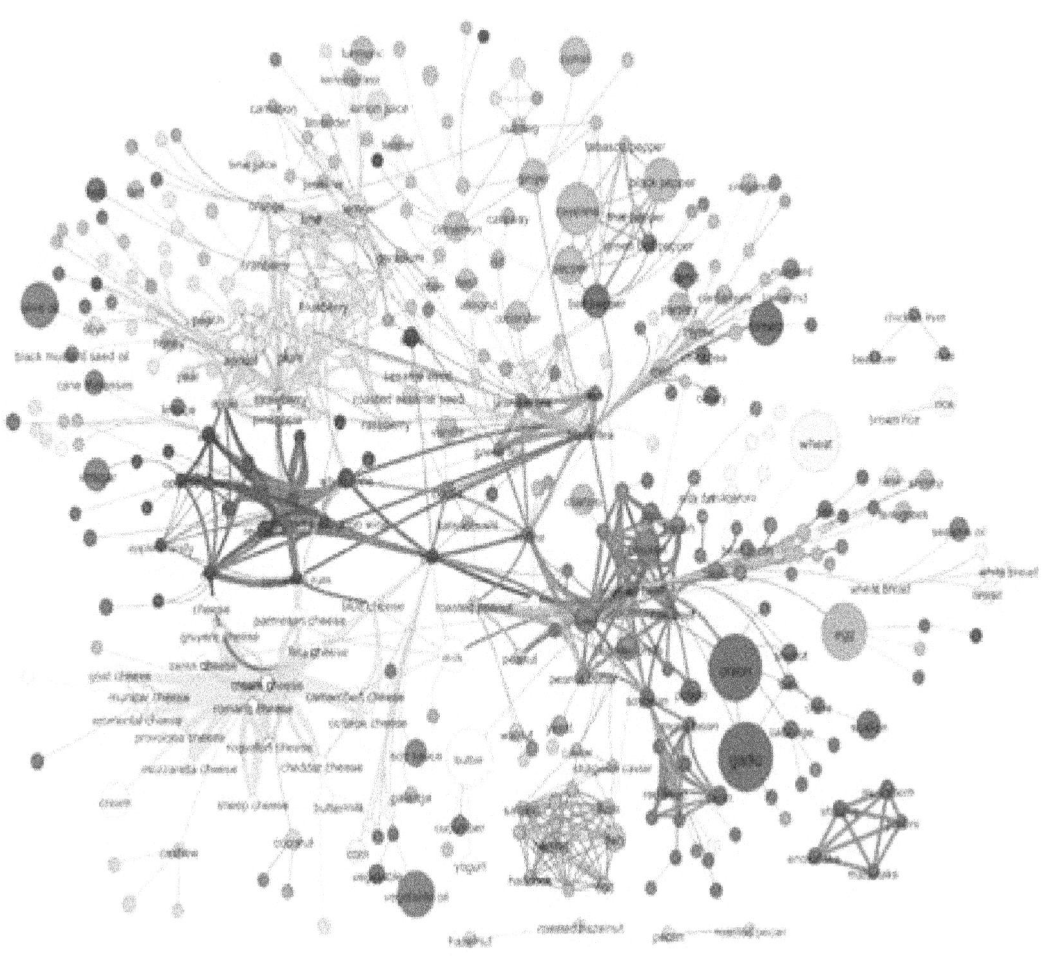

Figure 56. Source: "Vinography," a wine blog by Alder Yarrow.

Have you ever listened to several serious wine experts sharing their tasting notes about a wine that you yourself are tasting at the same time? I've had this experience several times, tasting wines with Robert Parker, Karen MacNeil, Andrea Robinson, Anthony Dias Blue, Frank Prial, and more. Without exception, not only are my tasting notes different from theirs, but all of theirs are different from each other. Some identify chestnuts, some taste tobacco, some cedar, and some espresso (espresso is a *method* of making coffee or even tea, not a taste). So, if the world's foremost wine experts can't even agree on what an individual wine tastes like in a controlled setting, how on earth could someone suggest they will know what it will taste like with rosemary-and-garlic-rubbed lamb shank with new potatoes and sautéed Swiss lamb? Are you detecting a bouquet of flimflam as I am?

Synesthesia: Derek Tastes of Earwax

What's my point?

Thank you, BBC.

The title of this section, "Derek Tastes of Earwax," is hard to get a handle on, which illustrates how odd synesthesia is in the real world. I could of course have said, "Does this coffee taste blue or number four to you?"

Synesthesia is not exactly a hot topic, which is why my major source of information has been an obscure publication by the Massachusetts Institute of Technology, known to most as just MIT. In case you are interested, it is craftily called "Synesthesia." The Derek-and-earwax reference is from a BBC *Horizon* program(me) on synesthesia. It is fun to watch. I guess it is time to define this term. Synesthesia is a joining or mixing of senses, especially sight, words, sounds, touch, smell, taste, and even proprioception. It is more common than you'd ever imagine. In its many and varied forms, it affects around one in twenty-three people, give or take. A synesthete may say that a flavor has them feeling that they are holding something. Here are some actual examples.

- Words and names have unique tastes, hence the name Derek tasting of earwax. Eight is an arrogant, fat lady, of course.

- Words can be crunchy, soggy, gritty, spicy, and purple.

- Describing a color can be intense for a synesthete. Non synesthetes have around five words for different greens. Synesthetes, fifty-four.

- Mayonnaise cannot taste pleasant because of the sound of the word.

- With tastes having colors, this makes things difficult for synesthetes to construct a pleasant food dish for themselves. There is an example where a man liked baked chicken with vanilla ice cream plus orange juice because of the tasty colors. By the way, it may not be obvious that pumpkin pie is a delicious purple addition to this combination. Don't forget that a large part of memory involves vision.

- Music in restaurants changes the taste of everything. To a synesthete, musicians' notes may have colors. There is an example of a classical conductor who did not like the bedroom in her new apartment until she realized that the colors were a flatted fifth (think fire engine siren).

- Some synesthetes can't read while they eat because the taste drowns out the words.

To add a personal note, my wife can't stand the words "chalk" and "pastels," because she experiences revolting teeth and mouth sensations and having to pee urgently. I don't understand either, and I am married to her. In every other way, she is perfect.

So, what is the takeaway? We have already established that we all live in our own flavor world, but this is taking it to the extreme. In the world of professional tasting (perhaps wine sommeliers are the best known), how much credence can we give to some "expert" trying to tell us what something tastes like? I can feel that I am making some enemies here. An ultra-high-class (read, expensive) restaurant in England offered a special evening for synesthetes. It was based on flavor, texture, color, smell, density, sound, light, and temperature. I am going to add "heavy on trigeminal perceptions." I bet you can't wait to experience this little treat but be prepared for a bill that will set you back around three thousand British pounds (sterling, not weight). The exchange rate is an inconsequential matter at times like this.

CHAPTER 7

ODDITIES GONE WILD!

What's my point?

Quirky flavor questions arise everywhere.

Androsterone

Are you a pig? If so, this is for you. Androsterone is a steroidal pheromone, but why should you care, unless you are a pig? A boar's saliva is a great source if you want some. Off you go. Call me when you have some. It does not work on humans, though it's a real turn-on for pigs—so much so that farmers spray it on pigs who are in heat. In his fabulous book *Entangled Life*, Merlin Sheldrake recounts how for years there was a logical theory that pigs' incredible ability to find truffles underground was linked to the truffles' production of androstenol. More recently, a study with stringent methodology cast doubt on this. This is an example of good science and demonstrating that it's acceptable to be found wrong in light of new information. Don't hesitate to get the book, by the way. Having said that, obviously the male readers among you can't wait to slap some under your armpits just in case I am wrong. Good luck. Even under ideal circumstances, you are only likely to turn ladies' heads away from you; however, a surgeon friend of mine said this might explain some of his earlier failures and successes with women. He is a good person, so I will keep him anonymous.

Androsterone is one of many chemicals that we interpret differently depending on what genes we have. Another example is exhibited by cilantro. Some perceive it as fresh and herbal and some offensively soapy. What I am getting at is that we are genetically programmed to like or dislike certain smells and, hence, flavors. On a purely cultural note, the word *coriander* (cilantro seed) is derived from the Greek word for bedbugs—you know, that delicious, irresistible, bug-infested-bedclothes smell.

Take a hundred people and give them a whiff of androsterone, and roughly a third will smell nothing, a third will detect something delicate like vanilla or wood, and a third will detect a strong smell of urine. No wonder we can't tell another person accurately what something smells like. We all live in our own flavor world.

Anosmia

What's my point?

Whoever introduced "anosmia" as a medical term did a sloppy job.

Anosmia is generally defined as a disease condition involving loss of smell. Put simplistically, "an-" = no, "-osmia" = smell. Therefore: no smell. That would be all fine, except that's not what anosmia is. Anosmia implies that some lucky sods have a sense of smell, and some have lost it. You either have it or you don't. Like good looks. It would seem to be hugely more appropriate to name the subject "smell and taste disorders," so that is what I am going to do. Taste is not smell. Just don't abbreviate it to STDs. From now on, there are smell and taste disorders and the A-word.

Any aspect of smell and taste perception can go wrong. By wrong, I don't mean something is missing, like losing your sense of smell. Oh, no—much worse than that. It often involves experiencing disturbing sensations that don't exist. They are phantoms. Parosmia, for example, is where you may smell a bucket of burning tires in your living room, but you know that there are no burning tires in your living room because you just checked. Disturbing. Sensing things that don't exist occurs within many medical conditions. Take synesthesia again. Synesthesia is where various senses get muddled up—get their wires crossed. You may hear colors and numbers. Sounds may induce nausea. A cruelly humorous example of this was exhibited by a physician's assistant in the operating room where I was working. The mere sound of the word "moist" would induce nausea in her. I will leave it to you to decide if we used to hang around her bringing the word "moist" into every sentence. I did say "cruelly humorous."

Loss of smell and taste encompasses a lengthy string of conditions mostly mirroring loss and grief when a loved one dies. How long is this string? No one knows. What do medical specialists say about it? They say squat, because they are as trained in smell and taste loss as the average man on the street corner. Doctors do not receive training in this. Well, of course they don't, you may say, because it is not rare enough to bother with, to which I would holler, "Is one in twenty common enough for you?" Divide the population of the world by twenty. If you had a bag of wheat grain for each of these, you could end world hunger and make yourself very popular in the third world. You don't believe me? Well, take your thoughts up with Doc Rob (ENT).

"Patients come to Ear, Nose, and Throat (ENT) doctors like me when they have nose problems. I may not be the first ENT doctor to admit that we know *very* little about our sense of smell, despite the importance it plays in our daily lives. Before COVID, and my nearly thirty years in clinical practice, I gained very little additional knowledge from what I learned in medical school! Much of that has changed since the pandemic. Smell and taste dysfunction was one of the first widespread symptoms identified in the early stages of the pandemic. Since then, research in this area has exploded. Before the pandemic, the most common cause of loss of smell (anosmia) in patients was idiopathic (translation: we don't know!). This was closely followed by post-viral infection, head trauma, nasal polyps, and neurodegenerative disorders such as Parkinson's disease and Alzheimer's, among others. Since the pandemic, post-COVID (viral) infection has taken center stage as the most common reason a patient walks into my clinic complaining of loss of smell and/or taste. There are different theories on how the COVID virus affects our ability to smell. Most researchers agree that it involves an inflammatory response targeted to the olfactory nerve, which forms a bulb between the top of the nasal cavity and the brain.

Based on data, I can reassure half my patients they will experience a full or nearly full recovery in less than sixty days. Another quarter will recover by six months, and the remainder by one year. Approximately 85 percent will improve overall. Unfortunately, this means around 15 percent will not experience a meaningful recovery. This is a sad situation for these patients, who too often suffer reduction in their quality of life as well as psychosocial maladies such as depression and even suicide. One of the most effective and proven treatments for anosmia, especially post-viral, is a practice called olfactory retraining. Olfactory retraining has been around for decades but gained new interest during the pandemic. Many variations of this technique have been described, usually with the patient exposing themselves to a few odorants one to two times a day for up to twelve weeks. The odorants can include lemon, eucalyptus, rose, and clover, but others can also be used, including coffee. The key is to have a distinct memory for the odorant and attempt to "remember" what it smells like during the process. Many clinical trials are ongoing to explore other treatment options, including pharmaceuticals.

The trigeminal nerve is another important nerve that we need to talk about. It is responsible for sensation of the face as well as the inside of the nose and mouth. It is also affected by COVID, as both the trigeminal nerve and the olfactory nerves share overlapping innervation. It is well known that anosmic patients can also experience a reduction in trigeminal function, including COVID-related anosmia. Why bring up the trigeminal nerve in this discussion? This gets back to Gerry. Among his many pursuits, which have included academic-level birdwatching, ultra-marathons, and coffee and beer judging, as well as our philanthropic work together doing medical missions in Ethiopia, Gerry is a born scientist and has made some interesting observations as it relates to anosmia and COVID. He has studied the topic of olfactory retraining to the point where I invited him to give a lecture to my staff. He emphasized nuances about the technique that can be easily missed, including the importance of odor memory, and addressing the psychological aspects for the patient. What about the trigeminal nerve?

While I have not been able to find any references in the medical literature, Gerry has proposed that stimulation of the trigeminal nerve can be additive to performing olfactory retraining. Focusing on the texture of food, how it feels in the mouth, and its crunchiness or temperature can help the whole system recover and restore some immediate pleasure from eating when taste and smell are reduced. The bottom line is that patients that experience prolonged taste or smell disturbance should not lose hope! While awaiting spontaneous recovery, they can take a proactive role using these techniques, and the vast majority will make a meaningful recovery."

This is Gerry, back again and please let me assure you that I did not pay Doc Rob for those kind words although I would have done. Thanks, Doc Rob. Enough about my fragile ego. Smell loss is just one part of a wider problem. It is like a bowel disease specialist only considering the first 20 cm of bowel and ignoring the rest. The smell-and-taste-disorder scenarios are massively widespread, and that's just by extrapolating from the pathetically little we already know. Given the foregoing, you cannot expect me to do due justice to the subject in this book, so I won't even try. I know what I am incapable of. (Never end a sentence with a preposition.) What I can do is explain the breadth of it: what is being done, why it matters, and the future of treatments, and generally scare the living daylights out of you. Job done by me.

Back in the real world, I am pleased to offer you the story of a patient and what he went through in his own words. Paul Joslin is my previous neighbor and someone who used to love to cook and eat. No more.

✳✳✳

"You don't die from what the medical community does know about your illness and disease; you die from what it doesn't know.

I was cruising along with my life in retirement, still strong and robust after a few rounds in the ring with prostate cancer. And with no—zero, nada—risks of heart disease, I was blessed with two stents. Not an issue. I could still proceed with my life plans for retirement now that my wife Sharon had retired.

Then, wham whammo—a sore throat that wouldn't go away. Biopsy…oh s**t, cancer. Not just throat cancer, but as a bonus, two minor shadowy spots on the lung. The battle is on, and it is fierce. Radiation on the throat was rough, and chemo was scary, and just before my final dose of radiation, I passed out on the living room floor, waking up in the VA Hospital with a feeding tube. Surely, I'm done with radiation. No. Doc says you need one more hit, then back to the machine for the final dose. At this stage, I had been using a feeding tube for several months—ergo, nutrition had no taste as it bypassed my mouth. But that's not totally true, because they want me to keep eating something so that my body doesn't forget how to swallow. I can still perceive flavor, and I consider myself one of the lucky ones, yet food doesn't meet my expectations, and texture seems to have attained greater importance in the overall flavor. (Gerry names this trigeminal nerve perceptions.) Devastatingly, the emotional desire to eat has gone. The joy of eating.

"I think I have a pretty good outlook on life and death. Life has been good to me, and death is not evil; it's simply the day we have been living for our entire lives. But I'm not ready to go yet, with a few more things left to do on my bucket list.

"When Gerry asked me to address 'taste'—or as this book suggests, flavor—for his research, I said, 'Why not?' Boy, taste, smell, speech, and eustachian tubes (not to mention the tube that gets the food from the mouth to the stomach) are complicated things. Then I must consider what is happening in the mouth, and that begins the process of getting the tasty food from the mouth through the throat to the stomach and beyond. Taste (flavor) is only part of the equation, and in the eyes of medical providers, not the most important thing. I guess keeping me alive was their priority. They seem to have no training as regards educating the patient on health, especially emotional health. They never spent any time, post op, discussing what changes I could expect as regards taste. However, I did have several conversations with Lauren, a speech language pathologist at the VA, but since the physical difficulties with swallowing and how to avoid aspirating my food took priority, taste was not a major issue.

"Let me describe for you my experience with taste on a timeline if you will. I had sixteen rounds of radiation therapy along with six infusions of chemotherapy. At the start of the procedure, around February 2022, I did not see any immediate change in my taste perception, but by the end of therapy, and after a brief hospitalization and insertion of a feeding tube, there were major changes. In my case, there was extreme sensitivity to some flavors. Salt tasted saltier and vinegary foods were almost intolerable, as were hot, spicy Mexican foods. The physical damage to my throat made it difficult to eat certain foods. Nuts, cookies, everything bagels, and even bread were difficult to swallow because my saliva no longer greased the throat as previously, making swallowing difficult and even dangerous for fear of aspirating food. Worst of all, Laphroaig, my favorite scotch whiskey, was so pungent, I could not bear to drink it.

"The war to recover my taste was on. At about this time, my guardian angel and health-care navigator hooked me up with someone who had the same procedure three years earlier. The big takeaway from my conversation with him was that 'food tastes different, so I simply tell myself that's what it tastes like now.' Not the most encouraging advice, but it made sense. Essentially, I had to learn to deal with it, to train myself to taste all over again.

"So now, over two years out, how am I doing? Flavors are not the same as they once were, but okay. Textures can be off-putting, especially things like nuts and cookies. Even a juicy steak can be tough to swallow. Almost every bite must be followed by water or another beverage to be sure to clear my throat and keep the airway open. I eat but seldom feel hungry and frequently have a take-it-or-leave-it attitude toward food. I rarely feel depressed, but when I do feel a little blue, my desire for food is suppressed. I recognize that a lot of it is in my head, so I eat because it's time to eat. Breakfast, what used to be my favorite meal, is a challenge, not because of flavor, but mostly because of texture. The silver lining is that all those no-no foods that you once tried to avoid because of calories are now free game. Wonderful, wonderful ice cream by the gallons, gobs, and gobs of whipped cream on desserts, and beer with dinner to wash down the food are all part of the protocol to regain weight and speed up recovery.

"Just recently, Kim, my guardian angel and health-care advocate, asked me to talk to a patient who recently had radiation on his throat and was losing weight big time. He was having difficulty with the decision to be made about a feeding tube. 'Food just tastes awful,' he said. 'I try to eat, but I just can't. I drink three to four Ensure a day but still lose weight. If I get a feeding tube, what can I expect? Does it hurt? What's it like to pour the liquid into the tube? Do you feel the liquid going into the stomach? How long do you have it in?' Mostly questions that can only truly be answered by someone that has had the experience. Everything else is hearsay. I said that if you can stand the taste of Jevity or TwoCal four times a day, then the tube may not be necessary. Otherwise, there is not much choice. We then talked briefly about tube removal, recovery, and taste before our conversation was interrupted—for a doctor's appointment, of course. Before we hung up, however, we agreed that he could call me anytime if he had questions or simply needed to talk. I sensed that it was important for him to know that there was someone to talk to if he wanted, even if he never picked up the phone.

> "Much of the struggle to return to normal is a psychological battle as well as a physical battle. I still eat spicy foods, just less spicy, and I pump up the spice occasionally to desensitize the taste buds. I tell people I am cauterizing my throat. As for Laphroaig whisky, alas, no more; a good Speyside or Highland single malt will do nicely, thank you."

***

What happens when you go to your doctor with smell loss? You may likely be referred to the nearest university Smell and Taste Center, where for many hundreds of dollars, they will run tests and tell you what you already know. You may be excused for asking, "Why bother?" to which I would reply, "Indeed." In the real world, one could skeptically ask why they would do this. The answer may be simple: they have funding to do it. I have struggled to find a more altruistic explanation but have failed. Feel free to educate me if I am wrong. I hope I am wrong. I will apologize as appropriate.

Treatment

Back to COVID-19. When people lose their sense of smell, it is primarily an emotional loss. The main symptom is depression. Doc Rob (ENT) engaged me to educate his team of doctors and others regarding what to tell patients on how to handle their smell loss. As previously explained, medical education does not include it. In my presentation, first off, I explained that COVID-19 smell loss is not their job. It is a psychotherapist's job. Sufferers should alert friends and family about their potential depression.

This is a huge subject but let me conclude by saying that patients might not be able to smell properly. Smelling employs cranial nerve one, but they still have cranial nerve five and taste buds. Their food and drink should be heavy on sweet, sour, salt, and bitter, plus stimulation of the trigeminal nerve with foods which are crispy, crunchy, smooth, hot, cold, spicy, and so on. I once said that if they have a pizza, they should also eat the box.

COVID-19 Smell Loss

Much has been revealed about COVID-19 and smell loss and possibly even more not understood. If one researches it all, there is one glaring takeaway: we really don't understand smell very well, let alone its relationship to COVID-19. Most people don't even know what the "19" stands for. No, I am not going to tell you. Writers—even science writers—get muddled between smell, taste, and flavor, as revealed ad nauseam throughout this book. Wine sommeliers will tell you that flavor is taste plus smell. I guess they will also tell you that cars are just metal and gasoline. True, but not helpful. Our main senses as far as the brain is concerned are sight, hearing, touch, and smell—arguably not in that order. Only smell, though, has actual physical molecules entering the body. Smell a banana and actual, real pieces of banana are entering your nose and stimulating the patch of tissue up between your eyes—the olfactory epithelium. This is the only brain tissue which, crudely put, hangs out in contact with the outside world. Yes, as mentioned earlier, BobHo—bit of brain hanging out. Doc Rob tells me, "Another reason nasal infections are so common."

Earlier, we established that smell is an evolutionary device to aid survival, not to enjoy wine and pizza. Dolphins can't smell because they don't need it to survive. Before we consume anything, using memory and emotion, we decide if it is a good idea or not. Burnt toast, rotting fish, excrement, or, in my case, cucumbers, do not aid our survival. All our likes and dislikes are learned.

It is 2023 now, but in 2019, I predicted that it would take at least two years for COVID-19 to settle down enough for normal socializing. The owner of the Great American Beer Competition got angry with me when I told him I believed he would get no income from it for at least two years. I was correct, and he is still angry with me. Clever, wasn't I? No, I wasn't. In 1994, the wonderful Laurie Garrett wrote a book called *The Coming Plague* wherein she predicted this whole calamity. 2019 was when I told those who would or would not listen that no one understands COVID-19. That is still true from my viewpoint. I guess this is my way of saying that I am not happy about making predictions and judgments regarding COVID smell-loss treatments. Please do not for a second think that this loss is trivial given that it accounts directly and indirectly for widespread social disruption and even death. That is twice in this book I haven't been funny.

There is no drug cure for smell loss. There is, however, a therapy that is used especially in England, and it is called "smell training." As I have laboriously emphasized in this book, flavor and smell involve memory and emotion in every instance. Smell training uses smells to try to jolt BobHo into retrieving memories, thereby revitalizing neural connections—that is, in theory. Essentially, the training involves using smell samples laying around and sniffing them as often as possible. Of course, you need to buy the special bottles of samples. Or you can just raid your kitchen and bathroom for things like lemon, vinegar, and perfume. That said, there is nothing to lose by trying smell training, and I would do it myself if I had smell loss.

People report degrees of success, but nothing happens quickly. Think in terms of weeks, months, and years. One might say that the successful ones would have recovered regardless. There is an old orthopedic saying, "Actively treat an acute condition and healing is likely in three months, but do nothing, and it might take ninety days."

Dysgeusia

Smell and taste disorders do not revolve simply around either having or not having the ability to smell or taste. There are a myriad of smell and taste perversions. Depressingly, some people smell burning rubber or even feces all the time. To emphasize the complexity of smell and taste disorders,

I have selected this one: dysgeusia, often defined as a disruption or disorder in the sense of taste. Symptoms of dysgeusia include a persistent rancid, foul, or salty taste in the mouth. There are a number of medications and conditions that can cause dysgeusia, including certain medications, cancer treatments, zinc deficiency, undiagnosed diabetes, poor oral hygiene, physiological changes like aging or pregnancy, and head injuries.

So why should you care? Well, if roughly one in twenty people have some version of a smell or taste disorder, that means among all the thousands or millions of beer, coffee, and wine judges, some of them can't be trusted, if you get my drift. The Specialty Coffee Association or the Beer Judge Certification programs have no process in which to ascertain if their members have a disorder. A friend of mine is a very well-known professional beer brewer, respected by all, yet he told me that now he can't smell properly, so he just drinks cheap, boxed wine. However, he still judges at competitions. It makes a mockery of the process, wouldn't you think?

Breast Milk

What's my point?

Mum's milk—evolution at its best.

Let's face it; we need some good sustenance after sliding down that slippery birth canal. It's been a bumpy nine months. Did you ever meet a newborn baby who turned her nose up at mother's milk? On the other hand, did you ever meet a newborn who could not get enough of bitter Brussel sprouts purée? The point of the foregoing is that we must learn to like or dislike everything we consume, but straight out of the womb, BobHo knows what is good without any experience. The newborn does not have much experience of flavors, so decisions on drinking are life and death. Literally. Bob Ho always says yes to mothers' milk. If the baby said no, then of course that would mean—well, you can work it out. It's not pretty. Not least, disappointed grandparents.

So, what does the baby know at this stage of its life? Not much. Just a huge subliminal interest in homeostasis. After all that swimming then sliding down the chute, a baby needs a drink. "Let me see, what shall I have? Something sweet to give me immediate energy so that I can cry and cry and cry. Hold on—I can smell it. It's just above me at the faucet. Mum should really get that checked out. It's leaky. Still, the smell made it easy for me to find. Lots of energy-giving sweetness. More please. I'm not ready for sour, bitter, and salt yet. I do remember the flavors of some of the foods Mum consumed during her last trimester though. I especially liked the carrots and garlic." Yes, the pregnant and breastfeeding mother can influence taste preferences in the baby by what she herself consumes. Taste is hardwired at birth. Flavor perception is learned. Smell is learned too—just ask any perfumer.

Considering the subject even more broadly, early nutrition has an enormous influence on a child's physiological function, immune system maturation, and cognitive development. In general, human breast milk is recognized as the gold standard for human infant nutrition. Breastfeeding is considered as an unequaled way of providing ideal food to the infant, which is required for his healthy growth and development. Human breast milk contains various macronutrients (carbohydrates, proteins, lipids, and vitamins) as well as numerous bioactive compounds and interactive elements (growth factors, hormones, cytokines, chemokines, and antimicrobial compounds). There are several new known metabolites of the breast milk and their potential role in infant development, such as oligosaccharides (especially for the baby's gut microbiome), nonprotein molecules containing nitrogen (creatine, amino acids, nucleotides, polyamines), and nonpolar lipids.

Considering the foregoing, one can see that formula to replace the real thing does not supply much more than a source of energy. That is not enough for the successful survival of the species, one could posit.

Castoreum

"I love your perfume."

"Thanks, I got it from the butthole of a beaver."

Forgive the language, but I don't imagine you are in the mood for a mammalogy anatomy lesson. Previously in this book, I mentioned that some smells/flavors at low sensory levels can be pleasant, but the opposite happens when the dose is upped. Castoreum is one of them. Castoreum is good in perfumes in part because it is a kind of yellow color. Not exactly puss, but you get the idea. Why, you may ask, did I include this? Because it is funny, and I know you too well.

Dogs' Paws Smell of Cheetos

I was asked about this, yet I had no idea regarding the answer, so I contacted my veterinarian friend Julia. Here's her educated response verbatim.

Hey Gerry,

Well, you do look smashing in a pink pussy hat, and yes, I know the smell well— variously described as Frito feet, popcorn feet, and the smell of corn nuts. Generally held to be due to low numbers of Malassezia yeast (normal flora on dogs) mixed with sweat-gland secretions (which dogs only have in any great number on the bottoms of their feet, between the toes). When dogs develop yeast overgrowth, the smell progresses to loaded Dorito. Interestingly, treating with antifungals such as ketoconazole makes the smell go away, but never entirely. Most dog owners grow to love the smell.

I guess the evolutionary question would be, why sebaceous glands on the bottoms of the feet? Sebum moisturizes the skin, protects it from infection, and increases its pliability/resilience. So, I would surmise that critters with sebaceous and sweat glands on the feet would have less infections/wounds and loss of full function than critters without. No tall dogs kick after eliminating, but all dogs have Cheeto feet.

Enjoy.

Good luck with the book and have fun—keep me up to date.

Julia

Geosmin

It occurs throughout nature, but you will know best what it is when I mention that smell of fresh rain on dry soil—pleasantly earthy. Soil bacteria produce it. You don't need many molecules to detect it. How many? I have no clue, but not many, by mathematicians' standards. Let's examine another chemical, indole. It has nothing to do with "indolent." Indole gives feces that distinctive, irresistible smell. Feces also contain hydrogen sulfide, which, oddly, is best detected at low levels. It is surely certain that scientists like to play with us when they state things like, "Humans can detect indole at five parts per trillion." What are we supposed to do with that, jump up and down and say, "Only five parts per trillion"? Most people can't define "trillion." There is not even agreement between the British and American definitions. The British trillion is a million times bigger than the American trillion, and there I was thinking that everything is bigger in America—don't tell the Texans.

I could not resist doing my own math. Imagine a fluid ounce (30 ml) of indole. If you diluted that in twenty-five thousand pounds of liquid, you could still smell it. Twenty-five thousand pounds is the equivalent of 104 1/2 Donald Trumps. You could dilute it down two million times and still detect it. Strong stuff, that indole. Yes, assertive. That is one reason not to put it in perfume, wouldn't you think? You'd be wrong though. It occurs widely in perfume and paperwhite daffodils. So, some smells are pleasant at low concentrations but decidedly not at higher concentrations. Putting that another way, indole and its relative skatole in low concentrations occur in ice cream, perfumes, jasmine, and orange blossom. At higher concentrations, it is the sneaky essential constituent of some military nonlethal weapons.

Corked wine is infamous and feared in the wine world. Commonly known as TCA or trichloroanisole, it does not occur in nature. It is the result of innocent fungi reacting with man-made chemicals used in disinfectants. You only need parts per trillion (British and American), but you soon adjust to the smell. However, time spent sitting in the expensive restaurant won't be long enough, I'm afraid. Once a winery gets infected, it can really do a number on profits due to lost sales. "Well, stop using those disinfectants and real cork," I hear you say. That's what they did. Other things can spoil wine, and the poor, old fungi still get the blame. Unfair. Funny thing: I tried to buy some trichloroanisole to taste test and could not find it anywhere. When you don't want it, you may get it, and when you do want it, you can't find it.

Rotten eggs. I put that in to get your attention because everyone knows what I am talking about. What is that rotten-egg smell? Hydrogen sulfide. My intention here is to confuse you. Apparently, low levels are offensive, but high levels are not. If you know what that means in real life, please contact me.

This next little snippet has me confused as to where to put it in this book, so I will just sneak it in here. What the heck are low, middle, and high notes unless you are playing a musical instrument?

Mocha

Ah, yes—that chocolatey coffee drink. The word *mocha* comes from the Arabic Yemeni port of Al Mokha, through where coffee beans were exported. Don't expect your cruise ship to drop you off there, because I understand that a sand bar closed off access many years ago. The green beans exported from Al Mokha were considered to have prized chocolatey notes when roasted and brewed. Now, if this character is so prized, the skeptic in me would inquire, "Why don't you just sprinkle cocoa powder on cheaper beans?" Think cappuccino. When you hear of a mocha food or drink item, don't get too excited about how exotic it sounds, because it just means cocoa powder has been sprinkled. Disappointing.

Farts and Farting

This section is not a case of me trying to get cheap laughs. In fact, my preference would be to not use the word "fart," however the only real alternative, "flatulence," is just a weak noun. Nobody ever flatulenced. "Fart" is a much better descriptor. Even *New Scientist* journal uses fart, admittedly in the Christmas issue. I have been called an annoying, old fart. I like that. So, "fart" and "farting" it is. Live with it. For me, the most interesting aspect is why we find farts and farting humorous.

Figure 57. "I could never understand why humans find farting funny."

It is not universal though. My interpreter Solomon in Ethiopia was horrified when I let one rip. To be honest, I am not sure that he had ever heard anyone openly (and joyously) fart before. Oh, wait a moment—here he is replying to my question on the subject.

"In our culture, we were grown up being advised that farting in the presence of people, mainly elders, is condemned. You must be careful and know where to fart and not. Especially females should take the necessary precaution not to fart in front of people. You remember when Mr. Sinclair farts in front me repeatedly? You know I was shocked and tried to keep quiet as though nothing happened. Anyway, this is what I feel and know about farting for the time being".

So let us get down to it (sorry, bad choice of words) and start with fart-aroma molecules and their source. The gas contains tiny amounts (define "amounts," Gerry) of sulfurous gases, notably hydrogen sulfide, the infamous rotten-eggs smell. This mingles with methanethiol (decomposing-vegetable smell) and the sweeter smell of dimethyl sulfide, prized at low levels in British ales.

The saving grace of all these off-putting attributes is the fact that we find farting funny. Ask anyone under 107 years old. In 1887, Joseph Pojol in France made his career by farting in front of theatergoers. He was a real fartist. Don't go thinking you can emulate him, because he trained himself to suck up air into his lower bowel in an opposite way that a trombonist will fill his/her cheeks and expel it. Joseph used his anal sphincter as a "mouthpiece." Come back, readers! I am not making this up, I promise.

Women produce less flatus than men, thanks probably to better self-control. In case you haven't gathered, researchers investigate these issues in all seriousness. They can't resist classifying everything, so it will come as no surprise to you to learn that farters occur in two varieties: smelly or inflammable. It still floors me that "flammable" means inflammable. A third of us are the latter, thanks to methane produced by our gut microbiome. Methane has no smell, but it makes up for it by catching fire at any given opportunity. No, I am not going to discuss fraternity students. We all produce a lot of gas. Is around five gallons daily enough for you? Once produced in your gut (and swallowed in food and drink), its destination is ill-defined—out of the mouth, reabsorbed, and farted.

On a prefinal comforting note, realize that if you smell a fart, know that it is actual, real molecules originating in feces. Kids love to hear that. Nuns, not so much.

On a final note, my mother was not known for her sense of humor, and she did not even know when she was funny. A famous phrase of hers was to comment on my capacity for being busy by saying, "Gerry, he's like a fart in a bottle."

Flavor Conservation

Flavor conservation is essentially the art of keeping food and drink tasting fresh. It goes way back for humans, but one man stands out as having had a massive influence on us even today: Louis Pasteur. Milk straight out of the cow: delicious. Leave it a week, and that all changes. Bugs invade it and chow down, producing flavor byproducts which many consumers don't enjoy. Louis Pasteur, in 1864, found that heating and cooling can interrupt this process, and today we still talk about pasteurized drinks.

Figure 58. *"I hope you used pasteurized milk!"*

Lactic acid is especially abhorrent to us unless we are educated to it. Probably the best examples are yogurt and Ethiopian injera flatbread. Americans call this process "spoiling" and the British "going off," as in, "Mum, the milk's gone off."

Fermenting foods and drinks can extend their gastronomic lives too. It's a huge subject, but mentioning yogurt, kefir, vinegar, cheese, kimchi, and sour beer, you get the idea.

Flavor-Wheel Foolery

What's my point?

Are flavor wheels just intimidating, arty hogwash?

Let me guess you are reading this with the book held relatively still except when you turn the pages. You are reading top to bottom, left to right. Can you improve on that for ease of consuming the written word? So, who were the chumps who decided that reading round in circles, bottom to top, upside down would be a REALLY good idea? Let's get practical and consider why the *New York Times* does not use this technique. Get my point?

Figure 59

Humans love to classify things. Linnaeus famously classified all living organisms, so I prefer to believe that classifying just tastes and smells would be a doddle for him. For the rest of us, not so much. In the 1950s, there was a hugely popular radio show which was based on a simple classification. Here is what Wikipedia says about it:

Animal, Vegetable, Mineral? was a popular television game show which ran from 1952 to 1959. [1] In the show, a panel of archeologists, art historians, and natural history experts were asked to identify interesting objects or artifacts from Britain and abroad, and other faculties, including university collections.[2][3]

The quiz show was presented by continuing a long history of bringing contributors to archaeology into the media limelight. Writing in 1953, the critic C.A. Lejeune described the show as having a "sound, full-bodied, vintage flavour".[4]

Let's look at it from another angle. If circles were a good idea, wouldn't advertisers be all over them? Taking it to the extreme...

Figure 60. "With Wells Fargo—was it a missed advertising opportunity?"

Flavor Wheel: Warning

This wheel is designed to intimidate you. Viewers must expect feelings of inadequacy and confusion.

Does anyone have any idea when and what the first flavor wheel was? Was it devised by someone whose head was spinning? Of course, the wine world loves them, but let's not stop with wine—coffee, beer, tea, honey, olive oil, cheese, tomatoes...shall I stop? Thinking that I could no longer be shocked by the next flavor wheel contender, I was wrong and shocked. You had better sit down for this next one.

The Body-Odor Wheel

The mind boggles. Please, no giggling regarding the fishy, cabbage, and boar taint items.

Figure 61

Glutton for punishment, are you? I'll leave it to you to research the urine wheel.

You could be forgiven for telling me to calm down, relax, and eat something flavorful. But flavor wheels matter a lot to some sectors of the flavor world. UC Davis in conjunction with World Coffee Research spent $200,000 developing their coffee flavor wheel and supportive lexicon. More depressing news on the lexicon coming later. Being a skeptic, I am forced to ask: what is it all for, and is it successful? I imagine you can picture me sniggering now. Let's dig deeper without going round in circles. Why do we need flavor wheels? They are an attempt to boil down the sensory elements of foods, drinks, and sweat (!) in a form that organizes and simplifies perception of (insert sniggering) the components. The wheels go from the middle outwards—general to the specific. They also go from the dubious to the nonsensical, in my curmudgeonly opinion. Let's dissect a flavor wheel. The following wine wheel in figure 62 is typical.

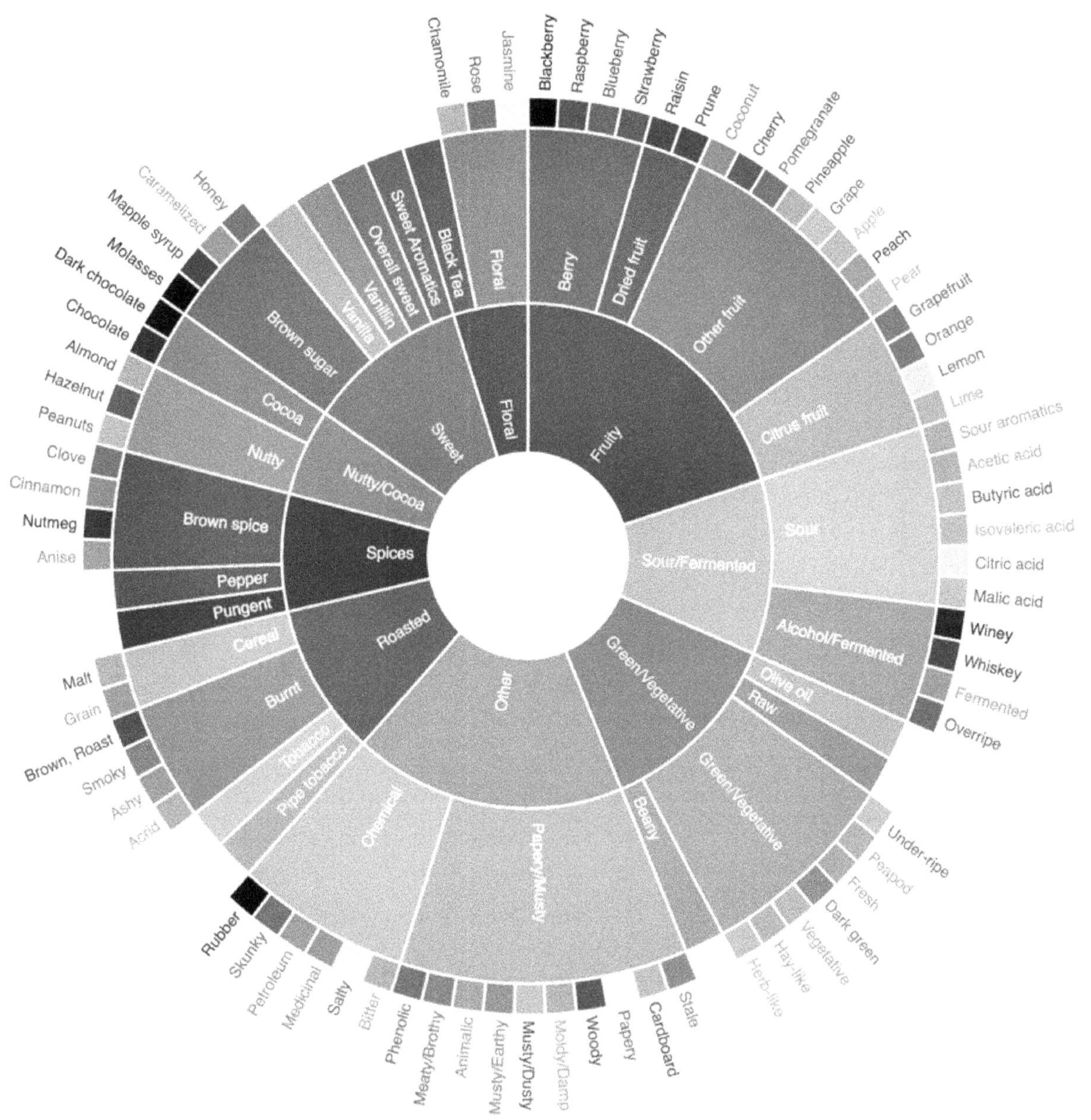

Figure 62

Pretty, isn't it? Useful? Not so much. In my experience, professionals and engaged amateurs don't use them on a day-to-day basis. It seems that they are just an educational tool. I have no problem with that. The new, $200,000 *World Coffee Research Sensory Lexicon* contains 111 (one hundred eleven) items. As just mentioned, it is the newbies who tend to use wheels, so let me reveal what they feel when presented with the coffee wheel: arty, intimidated, challenged, and out of pocket. What is the largest number of attributes it is possible to perceive? Answer: as many as you want because they are all in your imagination. If you say three hundred, then no one can prove you wrong.

Do you get the impression that flavor wheels annoy me? Correct. You know, I wouldn't mind, but the newbies are not told that they probably will never detect 90-odd percent of the items therein.

Lots of descriptors that people never use (or understand) on a daily basis pad out the wheel. Let me list a few from the wine one above.

Herbaceous/vegetative resinous ethanol

Chemical phenolic sulfa dioxide

Hot/cool pungent horsey

Oxidized mercaptan mousey (ugh?)

Other (?) sulfa dioxide orange blossom

Dried ethyl acetate diacetyl

Dodos (?)

Moldy cork

What the heck is "mousey"? Horsey—what part of the horse? The best one though, is chemical. They are ALL chemicals. The whole galaxy is chemicals. Show me something that is not chemical—apart from me laughing out loud. Worry not; at least it is intimidating.

More damned going round in circles: I have a friend who is less than half my age. He is so annoying though, because there seems to be nothing, he doesn't know details about. Christina and I have adopted him as what we refer to as "our nerd." He is a systems analyst for a major coffee company—bet you can't work out which one. Van the Man understands flavor perception better than anyone else I am acquainted with, and he enjoys mocking flavor wheels despite having one on his wall at work. "We never use it though." Proving the adage, "Great minds think alike," we came up with the concept that one flavor wheel should automatically lead on to another one and so on until it redefines itself at the original. "Is there a name for that concept, Van the Man?"

Of course he knew. "It's called a circular reference error and happens in the computer world, of which you know so little, old man Gerry." Yes, he actually said that. No respect. So, what does that mean to real, normal people? The wine flavor wheel has chocolate in it. Therefore, the chocolate item should be expanded into the chocolate flavor wheel, which has honey in it. The honey wheel has beer in it, and the beer wheel expands to include...wine. Voilà. Full circle. But we don't even know lemon from lime in the real world of Gerry's Kitchen Science Experiments.

Recently, I took the honey-tasting sensory course over two days. One of the attendees was a financial planner from New York City who was amazed at this new sensory world she never previously knew existed. She swooned at the presentation and almost woke me up. Is it just me, or do these wheels exist for the presenter to be able to say (to him or herself), "Look how smart I am. I understand all this, and you don't. What do you mean you don't pick up diacetyl?"

Now, here is the nasty truth, which my kitchen science experiment at MY PLACE revealed as just mentioned. Ask folk if they know lemon from lime, and they tend to say yes. No, they most likely don't. They don't know lemon from lime, but they are presumed to potentially detect one tucked into 111 items in a cup of coffee. Did I mention hogwash?

You are probably saying to yourself at this point, "Gerry, those last pages were a cloudy downer. Please lighten things up." What is a surefire way to brighten things up? Ask a coffee expert in sunny Hawaii. Who better than Shawn Steiman, PhD, Coffea consulting force and expert with books published on coffee? (www.coffeaconsulting.com) He kindly put down his cup of coffee and wrote the following for us.

What I want that section to say is, "These wheels are mostly—but not entirely—useless." Oh, yes, the wheel part is stupid. Someone could have made any number of versions of lists that are pretty and horizontal.

I fully agree that the wheels are mostly never used. IMO, they can be used in two useful ways (that maybe aren't worth $200,000). Of course, they rarely seem to be taught this way.

One, it introduces newbies to the possibilities. It may help them think outside their very tiny flavor box. They won't encounter most of the flavors (silly or otherwise) or even understand them. But it can help free them from constraint and give them power to really consider what they are tasting. Heck, folks often taste something but can't put a word to it. The terms on the wheel can help with that. Two, the idea that you can have a continuum of less specific to more specific is also freeing. If you can taste kumquat, great. If not, "citrus" will do. Even "fruit" is better than nothing.

The wheels are intimidating because experts want folks to think they can taste and remember all those terms. They should just be used as introductions to the process of tasting, to help folks be less intimidated.

So, yes, I like and agree with what you wrote. However, I think you should temper it with, "But they aren't a total waste of human invention."

Nose-Blindness: Olfactory Fatigue

Garlic? I particularly love it uncooked. Sometimes on awakening, my dear wife Christina will tell me she loves me, then kaboom! will tell me I stink of garlic. The smell is not something I can perceive though. Good, old fashioned garlic breath. Over the previous nine hours (guilty—I sleep easily and a lot), I have flooded my senses. Olfactory fatigue, nose-blindness.

It is not just garlic, of course, and it is theorized that a form of this is why we tend not to be able to smell our own farts. Workers in smelly environments become nose blind, as witnessed by my pool-chemicals supplier. She can't detect that pungent, intense chlorine smell soon after starting work in the morning.

What has this got to do with the subject of flavor, which is why you bought, borrowed, or stole this book? Perhaps it explains how we consume multicourse meals. We start simply with water and dry bread, progressing to intense deserts, coffee, and liqueurs. Why not the reverse?

People involved with competition blind tastings such as wine, beer, coffee, and honey tend to lose their edge as the tasting progresses. What is too hoppy at 9:00 a.m. may be perceived as pleasantly hoppy at 4:00 p.m., regardless of it being the same drink. As mentioned previously, we also calibrate to what we last consumed, not what we remember from previous items.

FMRIs: Your Brain on Food

If your brain had to choose one simple task to guarantee it lighting up like a Christmas tree, eating a French fry would probably do it. Humans have a few senses, which include sight, hearing, touch, pressure, smell, and taste—not in that order of importance. Each of them lights up areas in the brain. What does "lights up" mean? Should you have a hankering to know how your brain reacts to food, you can't just ask it. You can but expect sideways glances from those around you. You can't ask its owner. You can ask a physicist though. When physicists are not spinning their own heads thinking about dark energy and whether we and time really exist, some of them are using other forms of energy that we do understand. Enter the fMRI machine. Just to be clear, fMRI does not stand for "freakin' magical research instrument." No, it stands for "functional magnetic resonance imaging unit."

Not only does the brain have a network of nerves conducting electrical signals, but it also has a lot of blood vessels zooming hither and thither. So much thithering. Somewhat conveniently, it also happens that electrical activity mirrors blood flow, like the road that runs alongside the railway line. This means that to study brain activity, you do not have to look at the neural network at all. Like a certain hound, you just follow the blood. (By the way, did you adopt a dog as previously advised?) Undergoing fMRI examination has another advantage. You do not need a bowel prep or IV or urine sample. A valid credit card is handy though. The activity is represented by different colors. In a nutshell, stimulate the brain and follow the blood flow. Give you a nutshell, and nothing much happens. Give you a chocolate-covered nut, and bingo!

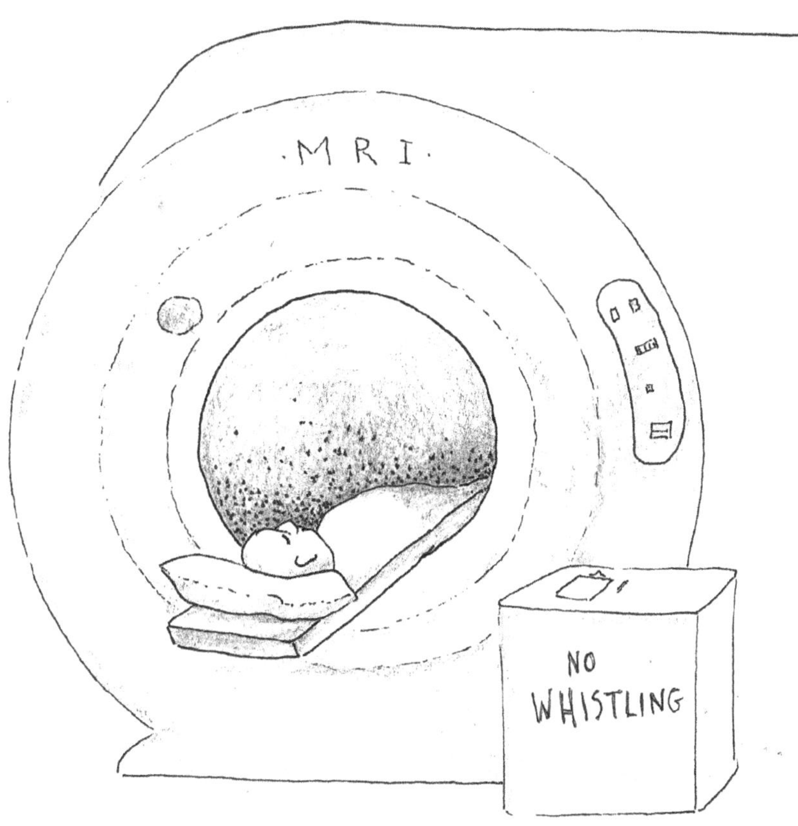

Figure 63. "Is all this really necessary? All I did was eat a banana!"

Gas Chromatography: The Parts of the Sum?

In medicine, we use CAT scans and MRIs. In flavor analysis, we use GC, gas chromatography, and mass spectrometry, MS. It is easy to go to the store and buy synthetic banana flavoring. Why? Because it is easy to make. Just one chemical makes anything taste like banana: isoamyl acetate. Rub it on your wife's finger, and it will taste of banana. Don't try this in public.

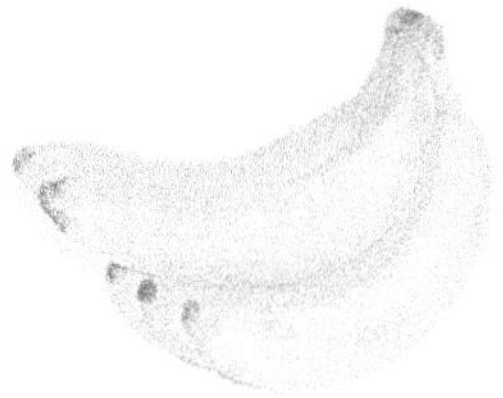

Figure 63. EASY.

So why not produce synthetic coffee? Or taking it to the extreme, a wine such as synthetic Château Lafite Rothschild 1999?

Figure 64. DIFFICULT.

Here is the answer. Whereas one can synthesize chemicals which remind us of banana or vanilla, products like coffee and wine are incredibly too complex to copy. Here's another stumbling block: just because a product contains a finite number of compounds, they do not occur in equal proportions, and, very importantly, humans are not equally sensitive to each compound. In fact, we are not capable of detecting some compounds. What does oxygen smell like, or even more dangerously, carbon monoxide? Put more simply, if a product contains a lot of compound A and a little of compound B, we may well detect compound B easily but A with some difficulty—the opposite of what one would think. Put even more simply, a little of minimal A may bowl us over, but a large amount of maximal B may pass us by. Some people may be able to easily detect compound A and others be unable to detect it at all. Yes, we all live in our own flavor worlds. Let's add that we may detect a compound but be unable to identify it by name. This is called having something on the tip of one's tongue. "I know what it is but just can't name it."

What is this gas chromatography machine thing? First thing to understand is that it does not come cheaply. Do not make plans to get a small loan from your mother-in-law and install one in your basement next to the washing machine. It, with the help of its operator, takes a sample of the item being analyzed and turns some of it into a gas, which can flow along tubes and reveal its hidden wonders. This is being horribly simplified for you, mainly because I don't understand it myself either. Using high-level chemistry and physics, eventually the gas chromatograph splits the sample up into separate gaseous compounds which it can identify and name chemically (as opposed to familiarly—Mary, Jim, and Einstein). As it does so, on a screen appears something that looks like this.

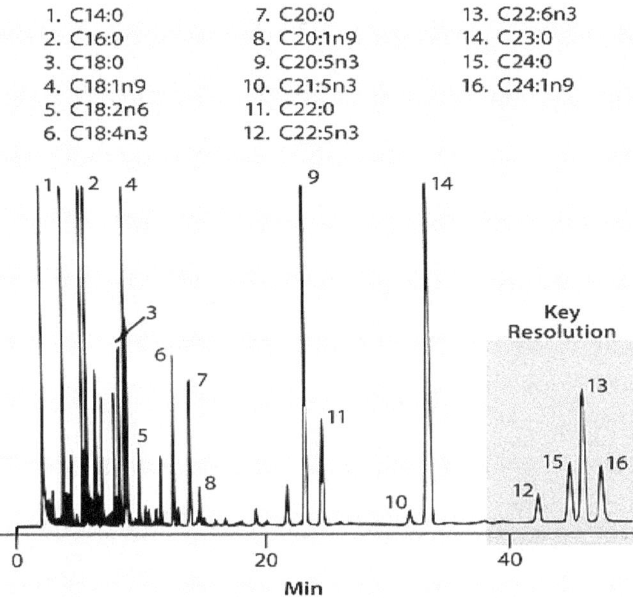

Figure 65

Each spike is a different compound. However, a short spike might smell more assertive to us than a whopping big spike. More is not necessarily better, in other words. The screen moves from right to left as new compounds are sensed. Here is where you come in. Well, me really, when I was "used" at a famous, international corporation laboratory.

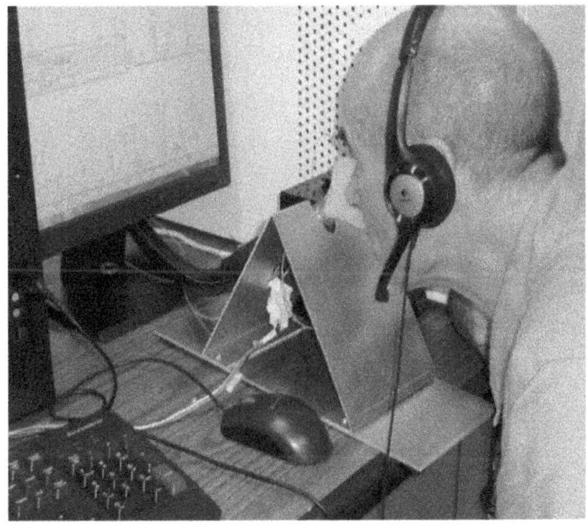

Figure 66

As each spike appeared, I verbally dictated into a recording device that was running in tandem with the screen. My task was to identify the smells, although there was no such thing as a wrong answer. Really exciting stuff, eh? Now, this being heavy-duty science, the little funnel in front of my nose has a scientific name. Go on, guess; I dare you. That's right, it is called—wait for it—a sniffer. In my case, we were analyzing coffee. My descriptors were mostly along the lines of "grassy, burnt rubber, pipe smoke." Nothing at all resembling coffee, and nothing vaguely pleasant. Here is the kicker then: coffee is made up of a collection of unpleasant smells, none of which reminds us of coffee, hence the title of this section, "The Parts of the Sum."

How is *Homo sapiens* supposed to combine hundreds of odd smells in the correct intensities to reproduce fresh-roasted coffee? We can't. In a nutshell, using the real-life coffee example, one may reveal and identify all the components of coffee but be unable to reconstruct them into coffee. So, was the experiment just a waste of time? You said it, not me. The parts are different from the sum of the parts. Don't even think about waiting for great old wines to be synthesized. Just pay the money, which may be slightly less than buying a gas chromatograph.

So, what is gas chromatography used for? The easiest to understand is to identify toxic gases in the air. These may include pesticides, combustion byproducts, forever chemicals, and good, old greenhouse gases, which are on course to bring us down sooner than governments want to admit to.

Unfortunately, I am led to the conclusion that gas chromatography has decidedly little to offer the understanding of flavor perception (see also "Corked Wine" ahead). An item broken down into a list of chemical names does not define it. It is just a list. It cannot define hate, love, or jealousy. When analyzing coffee or hops or whatever, as just mentioned, a sniffer is used—a person who sniffs and names the smells, as I did. Unfortunately, no one ever tests the sniffer to see if he or she has a smell or taste disorder. Let us imagine that the GC corporation used a PhD chemist who is famous as a beer writer, brewer, and personality to do some sniffing. Hops were the subject. Surely, he would be an ideal volunteer. Yesterday, I challenged him to identify what a drink sample was. He identified it as tea. It was in fact a mix of pomegranate, apple, and raspberry juices. The corporation which I had assisted used a sniffer for its hops project who was known to have a smell/taste disorder because he told me he did. What is the phrase? Garbage in, garbage out. (That reminds me, I must take the...)

Hot Peppers at the Bar

Barman (leaning across the bar with a soft, questioning voice): "I have customers who want hot peppers in their drinks. Do they know something that I don't?"

Gerry (with a wry grin and a dry gin): "Yes, they do. They know what they want, or at least what they think they want. You're an intelligent man, so let me explain a bit about hot peppers and humans. You can do anything you want with hot peppers, including putting them where there is a decided lack of sunshine, but it does not mean that it is a good idea." (The barman laughs openly.)

"The tongue does many interesting things, including that which you are giggling about now." (The barman does something with his tongue which only humans can do.) "We are the only mammals with the dexterity to do more than just stick our tongue in and out. With great dexterity, it pushes the bolus of food round the mouth. Push the meat to one side, and you double the pressure you can exert on it.

"Apart from famously detecting sweet, salt, sour, and bitter, the tongue is also loaded with touch and pain receptors—the same pain receptors which make hot peppers hot to us. Heat receptors on the tongue tell us if our coffee is too hot, ice cream gives us a headache, or a cigarette magic trick has gone horribly wrong. This burning sensation is your body's way of saying, 'Spit it out immediately, you idiot.' Evolution knows. Survival again. No one is born liking hot peppers. It is all learned behavior. We are born liking mother's milk and not much else."

Barman (eyes wide): "So are you saying that eating hot peppers and masochism have something in common?"

Gerry (eyes wider): "My goodness, you are ahead of me. Ultimately, whatever the customers want is good. We are not to judge, apart from the level of pleasure we are enabling."

Barman (with a slight flush to his smooth skin): "I have just eaten a jalapeño. How do I dampen the burn?"

Gerry (with a cruel grin): "There is only one surefire—sorry about that choice of word—remedy. Wait until tomorrow. The good news is that the burn itself will not harm your mouth tissue, unlike hot water, but your reaction to it might cause that little problem called reversible cerebral vasoconstriction syndrome, typified by the worst headache imaginable. It is as painful as it is uncommon. Of course, if you vomit, it has been known for the consumer to rip their esophagus. Not exactly trivial, nor is it the kind of heart attack that killed a man. But on the bright side, the Chinese studied 480,000 consumers, and those who ate chili pepper daily were 14 percent less likely to die than those who ate them less often."

Barman: "Is there any chance you won't return here?"

Lemon and Lime, Always Fine

What's my point?

Hold the lime; lemon is fine.

What's classier, lemon or lime?

What's more expensive, lemon or lime?

If you want to make the best citric statement in a drink, do you choose lime over lemon? Lime, probably. Experts and classy home cooks generally prefer lime. Can humans tell the difference? No, to which bartenders, chefs, flavor expert professionals, and my wife respond, "Rubbish. I can tell the difference!"

Put samples of lemon and lime juice side by side and ask which one is lime. They will say, "This one is lime," and they will probably be correct. Good old deductive reasoning narrows the choice. If, however, I do a real (by that, I mean a Gerry-style) blind test, they can't. If out of the blue, with no build-up, I squirt a few milliliters of liquid into your mouth and ask you to identify it, you can't. Part of my test is this condition: you get it wrong, and you pay me ten dollars. I find that this halts the guesswork. Don't give me that look! You may well still be confident in your ability. On two continents (no, not including Antarctica), I have tested flavor professionals, and not one of them has been positive that they can identify the sample.

The next piece of fun up my sleeve is to squirt a liquid which is a fifty-fifty mix of lemon and lime. Guess. Let me make it easier; flavor experts I have tested can't identify lemon or orange in a fifty-fifty mix. The takeaway: use the cheapest lemons you can buy, and if I come at you with a squirty thing, back off.

P.S.: In the wine sommelier world, when they are having difficulty describing a wine, they have a fallback: "Lemon and lime, always fine," because they know it is hard to argue with these two descriptors. Then, add on, "Delicate notes of this and that." My freethinking, wine-expert friend Mitch Ancona added that one can hedge one's bets and use, "Apple and pear—always there!"

Gin Botanicals

What's my point?

Just because you can doesn't mean you should. One could say that vodka is gin without the fruit and vegetable. Do you know what you can put in gin? Here's a clue: it is an eight-letter word beginning with A.

Anything.

It didn't start that way decades ago in Holland and Britain and many other countries. They were simpler times, when an infusion of juniper berries (which are not berries) did the trick to separate it from vodka or other hard spirits. Nowadays, anything goes—a marketer's dream. To be honest, it is not just gin developers who try to gain a market edge by making their product more and more and more complex. If a gin distiller has a gin with ten ingredients, does he/she do blind tests to see whether if any one ingredient is removed, it can be perceived?

(Do you remember 1 + 1 + 1 + 1 + 1 = 1 or less?)

Gin developer Robert Cassell from Connacht Whiskey Company in Ireland told me confidentially, "They are putting too much stuff in gin." Good Irish sensibility.

Palate

What's my point?

A near-useless word unless you are having buccal cavity surgery.

The chances are strong that you have never heard a parent say, "My two-year-old has a terrific palate," or, "My eighteen-month-old baby is going to be a great, race car driver." Why not? Both skills are learned, not inherited. That tells us much about acquiring a discerning palate in that anyone can do it if they put the effort into it. The same is not true regarding becoming a race car driver, in that other aspects like fear and perhaps recklessness come into it. Anatomically, your palate is the tissue on the roof of your mouth, which is enervated largely by the trigeminal nerves (see "The Mouth: Trigeminal Perceptions," chapter 6).

Developing a "good palate" for the average person requires an interest in eating and drinking. In practice, this means an interest in identifying and describing flavors (not smells). You would need to think about what and how you eat and what you choose to eat and not eat. As we get older—especially over eighty—we tend to lose some ability to recognize smells and therefore flavors, although we can still retain an interest with memories of what we have eaten in the past.

People often use the word *palate* meaning "sense of taste." More accurately, they should invoke Gerry's oft-repeated trigeminal perceptions, described by them as aftertaste, smooth, chalky, resinous, and my favorite, skanky.

In the drink-judging world of wine and beer, the judges' tables usually have a dish of palate-cleansers. That's a technical way of saying white bread and corn chips all washed down with water. Of course, these will do little to remove bitter tastes and astringency and—still, as long as they are having fun.

Pheromones

Mention of the word *pheromone* leads most people's thoughts to attracting mates. The following is for sale: Love Scent's Super Primal Pheromone Oil. It's a sex thing. Pheromones are chemical messengers and are capable of delivering nasty as well as nice information. Sorry, but as of writing, humans have no sex pheromones. That is not to say that they won't be found. I think it is highly likely that they will. Just not today. Certainly, humans can be attracted or otherwise by another's smell, but if you think that dabbing on some Love Scent Oil will bring them running, think again. Body odors can attract or repel. One excuse that does not work on a date is, "I'm sorry, my dear; I don't have BO, just a very nuanced pheromone colony." It's a big subject and not really within the scope of this book, but just to put it in perspective, dead bee larvae put out pheromones to attract live bees to cart them away from the hive. Bring out your dead! That's not the stuff of *Playboy* magazine.

There are opinions that we produce fear-omones—chemicals in sweat expressing fear. Beware the smell of fear.

Supertasters

Are supertasters born? Solely on a reproductive level, the answer is yes. What do top tasters, perfumers, skateboarders, Olympic athletes, concert pianists, jooking performers, and rap singers all have in common? Come on, out with it. They all got where they are by practice, practice, practice. None of them woke up one day at the top of their game. Certainly, there is genetic variation—natural tendencies—much like some of us are taller and balder than others. There are examples of very young children playing musical instruments. Unstoppable. Therein perhaps lies the answer. You can bet this child spends massive numbers of hours at the instrument. Perfumers train themselves, as do coffee, wine, and beer tasters and skateboarders. Regardless, it appears that some are just better than others—naturally better. However, what they all have in common is not that they have some magical, in-born, exceptional ability bestowed on them at birth, but that they are all intensely interested in their quest. Someone who is not at all interested in music is unlikely to end up in Carnegie Hall unless as a cleaner.

In the *New Scientist* journal, May 19, 2018, page 42, there is a review of the book *Superhuman: Life at the Extremes of Our Capacity* by Rowan Hooper. A quote regarding extraordinary people: "It's never genes or environment; it is always both things together." Genes strongly influence the drive to be an exceptional taster or smeller. What was that regarding practice, practice, practice? Allow me to put it this way: "So-and-so has an amazing passion and practiced interest in tasting and smelling."

The concept of "supertaster" was coined by Linda Bartoshuk. There is much about her on the internet, so there is no need for me to rehash that information. Polite, entertaining video. She proffered that people with a high density of taste buds (especially bitter receptors) on their tongue make better tasters, special. At a conference, I had breakfast with her and expressed my flimflam opinion of her theory. Her scientific answer was to continue, head down, eating her eggs. Apart from that, I like her a lot. According to Linda, you are at an increased likelihood of being a supertaster if have inherited a dominant variant of the TAS2R38 gene from your parents. So, have you? You can buy supertaster test strips, which are very handy for letting you know if you have more bitter receptors on your tongue than the person next to you. Fascinating, isn't it? You can also shortcut that by looking in the mirror at your tongue.

If you are sensitive to bitter, then you may be similarly sensitive to sweet, sour, and salt.

According to the bitterness sensitivity strips, I am a supertaster. As far as I am concerned though, it just means I tend to experience taste in neon. A little really shines. So, salt makes me cough. I enjoy bland more than intense. A little pepper is enough. I hate cucumbers with a passion. Same for watermelon. You could label me as being gastronomically boring to be around. In fact, I would agree with you. I have little interest in eating despite having written this book. The low tasters tend to be more likely to enjoy fatty, sweet foods. They also tend to be bodily fatter. They like brassica vegetables, hot peppers, spicy foods, highly roasted coffee, and hoppy beer. Realize that I am aware that these are sweeping generalizations. There are degrees in everything, including middle tasters. What the heck does it matter?

Taste Panels

Big dollops of intimidation? Consumers are widely offered recommendations on drinks by individuals and panels of people who are trusted to know what they are talking about. Wine, whiskey, coffee, and beer again come readily to mind. There are all manner of competitions and tastings which guide the followers to what is good, better, and best—and worst. The general public is mostly excluded from these events, letting the professionals prevail. Let's take a real example. You may say, "Lucky you," when I tell you that I was chief steward at the Great International Beer, Cider, Mead, and Sake Competition. It's for professional entrants only. Thousands of bottles were under my eagle, vigilant, protective eye.

Figure 67

Figures 68 and 69

The drinks were divided into logical categories. Obviously, beer was separated from sake, which was separated from mead and cider. Within the beer category, there were dozens of subcategories. Stouts, barley wines, pilseners, weizenbiers, smoked beers, and so on. Two or more judges were assigned to each flight. The panel's job is to find the best three drinks in their flight as well as offering advice and comments to the brewers. Winners can use their glory in their advertising to proudly say something like, "Gold-medal winner."

This is all very well, but what happens if a judge simply does not enjoy the drink she is judging? That is why judges are trained to put their personal feelings aside. All smoked drinks are abhorrent to me, but I am still able to give a well-brewed example a high score. Yes— "Beautiful example of the smoked style. Hated it."

A similar scenario plays out with coffee, wine, honey, tea, olive oil, and so on. Regardless of the product, judging forms follow a broad pattern of assessing aroma, appearance, flavor, mouthfeel, and overall impression. Coffee wants more information on acidity, whereas wine is interested in astringency and alcohol, and so on. So many marks for this, so many for that; add them up and find the winner. In practice, judges may disagree with their own weighting of scores and fudge them until they reflect what their gut tells them. In other words, they cheat a bit. We all do it all the time, so sue me.

So how good are the judges? Years of experience have taught them to independently judge one drink from the rest—you would think. If only it were that simple. Craig Shelton of King's Row Coffee tell us that "our palate measures all taste relative to what immediately preceded it and in relation to our environment." That means the drink we are judging/rating now is in fact being mostly compared to whatever we consumed just before it. The item before is our reference point, not the preceding years of experience.

BEER SCORESHEET

http://www.bjcp.org   AHA/BJCP Sanctioned Competition Program   http://www.homebrewersassociation.org

Judge Name (print) _______________

Judge BJCP ID _______________

Judge Email _______________
Use Avery label #5760

Category # ______   Subcategory (a-f) ______   Entry # [____]

Subcategory (spell out) _______________

Special Ingredients: _______________

Bottle Inspection: ☐ Appropriate size, cap, fill level, label removal, etc.

Comments _______________

BJCP Rank or Status:

☐ Apprentice   ☐ Recognized   ☐ Certified
☐ National   ☐ Master   ☐ Grand Master
☐ Honorary Master   ☐ Honorary GM   ☐ Mead Judge
☐ Provisional Judge   ☐ Rank Pending   ☐ Cider Judge

Non-BJCP Qualifications:

☐ Professional Brewer   ☐ Beer Sommelier   ☐ GABF/WHC
☐ Certified Cicerone   ☐ Adv. Cicerone   ☐ Master Cicerone
☐ Sensory Training   ☐ Other _______________

Descriptor Definitions (Mark all that apply):

☐ **Acetaldehyde** – Green apple-like aroma and flavor.

☐ **Alcoholic** – The aroma, flavor, and warming effect of ethanol and higher alcohols. Sometimes described as hot.

☐ **Astringent** – Puckering, lingering harshness and/or dryness in the finish/aftertaste; harsh graininess, huskiness.

☐ **Diacetyl** – Artificial butter, butterscotch, or toffee aroma and flavor. Sometimes perceived as a slickness on the tongue.

☐ **DMS (dimethyl sulfide)** – At low levels a sweet, cooked or canned corn-like aroma and flavor.

☐ **Estery** – Aroma and/or flavor of any ester (fruits, fruit flavorings, or roses).

☐ **Grassy** – Aroma/flavor of fresh-cut grass or green leaves.

☐ **Light-Struck** – Similar to the aroma of a skunk.

☐ **Metallic** – Tinny, coiny, copper, iron, or blood-like flavor.

☐ **Musty** – Stale, musty, or moldy aromas/flavors.

☐ **Oxidized** – Any one or combination of stale, winy/vinous, cardboard, papery, or sherry-like aromas and flavors.

☐ **Phenolic** – Spicy (clove, pepper), smoky, plastic, plastic adhesive strip, and/or medicinal (chlorophenolic).

☐ **Solvent** – Aromas and flavors of higher alcohols (fusel alcohols). Similar to acetone or lacquer thinner aromas.

☐ **Sour/Acidic** – Tartness in aroma and flavor. Can be sharp and clean (lactic acid), or vinegar-like (acetic acid).

☐ **Sulfur** – The aroma of rotten eggs or burning matches.

☐ **Vegetal** – Cooked, canned, or rotten vegetable aroma and flavor (cabbage, onion, celery, asparagus, etc.)

☐ **Yeasty** – A bready, sulfury or yeast-like aroma or flavor.

Aroma (as appropriate for style) ___ /12
Comment on malt, hops, esters, and other aromatics
____

Appearance (as appropriate for style) ___ /3
Comment on color, clarity, and head (retention, color, and texture)
____

Flavor (as appropriate for style) ___ /20
Comment on malt, hops, fermentation characteristics, balance, finish/aftertaste, and other flavor characteristics
____

Mouthfeel (as appropriate for style) ___ /5
Comment on body, carbonation, warmth, creaminess, astringency, and other palate sensations
____

Overall Impression ___ /10
Comment on overall drinking pleasure associated with entry; give suggestions for improvement
____

Total ___ /50

Outstanding	(45 - 50): World-class example of style.		
Excellent	(38 - 44): Exemplifies style well, requires minor fine-tuning.		
Very Good	(30 - 37): Generally within style parameters, some minor flaws.		
Good	(21 - 29): Misses the mark on style and/or minor flaws.		
Fair	(14 - 20): Off flavors/aromas or major style deficiencies. Unpleasant.		
Problematic	(00 - 13): Major off flavors and aromas dominate. Hard to drink.		

SCORING GUIDE

	Stylistic Accuracy					
Classic Example	☐	☐	☐	☐	☐	Not to Style
	Technical Merit					
Flawless	☐	☐	☐	☐	☐	Significant Flaws
	Intangibles					
Wonderful	☐	☐	☐	☐	☐	Lifeless

BJCP Beer Scoresheet   Copyright © 2017 Beer Judge Certification Program   rev. 170612   *Please send any comments to Comp-Director@BJCP.org*

Figure 70. I challenge you to analyze this without laughing. Yes, it is ultimately just beer.

There are famous, supposed wine experts who rate wines and give them scores, which you see on little cards on the racks in wine stores. A wine with a score of ninety-three must be better than a wine with a score of ninety-one, one would think. Call me crazy, but wouldn't one assume that these "names" would tend to agree with one another and use the same descriptors for any given wine? As you have probably guessed, this does not happen much in practice. One expert uses a different vocabulary from her counterpart. Does that mean that expert X is unable to describe a wine to expert Y? So that is their little secret then. That little card in the wine store is useful only to one person: herself. If you buy a wine with fancy descriptors attached, don't expect to find any of those attributes when you sample it. There is a well-known wine sommelier who in a YouTube tutorial tasted a wine and followed with, "I get dark fruit...notes of plum—no, blackberry—no, black raspberry—no, red raspberry." Okay, lady; just take us on a trip down the fruit aisle. The thing is that she could say anything she liked, and you could never prove her wrong. I sometimes say that something tastes of gold, frankincense, and myrrh just to watch the faces around me. If that does not do it, then I try goblins. One thing is well known though: just mention a flavor or smell, and those around you will probably detect it too. Somewhat cruelly, I like to offer, "Notes of apricot?" and wait for the agreement. It works the same for peach.

Let's look at it in another way. I can describe an animal, and it may be clear to me what the animal is that I am describing, but my description might be useless to someone else. They might interpret my description very differently from how I imagined. Let me ask you, what is this creature? Sit down.

Yes, this is a test:

- It is a mammal.
- It is mostly all gray.
- It has a short, somewhat insignificant tail.
- Its ears are rather distinctive.

To me, it is an elephant, but to Mary, it is a mouse. We are both correct. In the same way, a famous wine writer may describe a wine which fails to guide another famous wine writer as to what it is.

At this point, it might behoove you to go back to the chapter on the Laing and Nicholls limits (chapter 6). Why bother with all the waxing lyrical when tasting? The answer: because it's fun, and as Professor Gordon Shepherd, who wrote *Neuroenology*, told me, "And because it brings a focus to the tasting process." The wine industry (and many other industries) is keen on securing our positive emotions. It's all show business, folks! I love it too and also fall for it. It is mostly innocent, so why not have fun? The odd thing is that when I talk with wine professionals about it, their first response is to laugh. Yes, laugh. A wine retailer friend of mine believes he could get rid of most of his wine selection, and no one would miss much except the overfull shelves. At the BevCon 2017 beverage conference, I offered some wine sommeliers a drink which was heavily spiked with vinegar. None of them picked up on it. They laughed when I exposed them. Even more so, it amused the Kentucky whisky distillers who were present. These folk know how to laugh.

Terroir: You Really Want Me to Define It?

Have you got a few spare minutes to waste? Look up *terroir* in dictionaries as well as whiskey and wine information sources. Let me guess, now you really can't define it. Here is my version: terroir is something other or not to do with taste. Especially the soil. Possibly.

The subject of minerals arises frequently. Minerals are all solid, except mercury. No, you can't have a mercury sword. Terroir expresses itself in flavor, which utilizes smell, so minerals play no direct contribution to terroir. You can't smell tin. The soil is a vehicle to hold minerals, but more importantly and numerously, trillions of microbes in each handful (even if you have a small hand). Microbes can produce smells. If a wine buff tells you a wine has mineral notes, walk away while laughing your head off. They love that—I've done it.

Cast of Trillions

Back in 2019, thedrinksbusiness.com reviewed a study that showed terroir does influence whiskey. Whiskey with an E refers to Ireland. The article opened with a gorgeous photo of barley swaying in the breeze—except that was wheat. That did not inspire in me much hope for the rest of what I was about to read. Essentially, they found that the environment affected flavor produced by the grain. They took a full year to find that out.

> Definition by them: Terroir is the 3D influence between microclimate, soil, and place on the plant.

> Definition by me: The foregoing is highfalutin mumbo jumbo/twaddle.

The grain samples varied in the levels of selenium, chromium tin, barium, cadmium, zinc copper, and aluminum, none of which humans can detect in food and drink. Was that simply pointless or not?

"New made spirit" (?) was highest in furfural, which elicits notes of bread. You mean, just like Maillard! I bet you are always going on about furfural in your food and drink. "Darling, can we have less furfural on the pizza next time?"

Are you ready for the big summation? Wait for it. Different whiskeys are different. Well, knock me down with a bucketful of furfural. Perhaps we should take a short moment to explain terroir that may be useful. Terroir is the influence on flavor produced in substantial part by the soil microbiome. Why didn't they just say that? I hope they did not get paid.

French Fries

What's my point?

Hey tubby—want another French fry?

Ah yes, French fries—Christmas for the brain. Forgive me for appearing rude, but we don't normally associate French fries with slimming, do we? French fries, or chips in the UK. Chips in the UK are crisps in the UK. Confused? Let's call them "fries" hereafter. World War I, roughly 1920, is where it started. Returning soldiers to the UK talked about this wonderful food item in France, fried potatoes from France that they cleverly named French fries. Did those soldiers for a second imagine what a huge, international culinary delight they would become? Twenty-five percent of the vegetables consumed in the USA are potatoes. A medium portion of fries provides 25 percent of a day's energy requirements. The extra can be packed round your liver if you wish—very approximately, of course. Much of what I am writing about fries is mimicked by pizza, I know, so leave me alone. Both have a carbohydrate base acting as a vehicle for toppings and flavor additives. Let's put the pizza back in the box for now and just discuss fries. They have it (just about) all, and what they don't have can probably be piled on top or at the side.

A day in the life of a fry.

Put it in your mouth and chew. Don't swallow it whole, for several reasons, not least to maintain peace in the home. A trigeminal nerve circus now explodes into play. Springiness indicates freshness and proper cooking degree. Crispy, crunchy on the outside, contrasting with smooth, soft, creamy inside. Hot too. If it was cold, you would spit it out. Now that you are relaxed, here is a true Gerry story:

I was once addressing some top regional wine sommeliers (identified by their two-hundred-dollar bottles of wine in cloth bags). I placed a fry on a napkin in front of each of them. "After the count of three, eat." In their mouths they went, and out they came immediately, faster than escaped rats out of a research lab. The sommeliers were expecting warm and crispy. Memory and emotion and expectation played their parts again. They were expecting crispy and hot but got cold and limp as a wrist. Trigeminally speaking, fries tick all the boxes. Next up: taste (sweet, salt, sour, and bitter).

Sweet

Sweet in fries comes from starches converted to simpler sugars. Thanks to the Maillard reaction (same for bread, pizza crust, bacon, etc.), potatoes are converted to something sweet. Our love of sweet starts with mother's milk.

Salt

You can sprinkle it on, but there will be some in the ketchup if you use it. Salt enhances flavors despite having no smell itself. What about table-sugar smell? Zip. We like salt. It comes from the word *salary*. We also like money. Get a plain peanut and eat it. Stop there. Easy enough. Get a peanut and add salt. Stop there. You can't, can you?

Wait; they are coming up. The mouth likes all this trigeminal nerve stimulation and wants more, so we obey by chewing. Mastication pushes the bolus of food round the cheeks and coats our teeth. All that wonderful flavor we are perceiving in the mouth, eh? Wrong. It is all concocted in the brain. You nearly forgot about the two other tastes, didn't you?

Sour
Easy: add vinegar if the vinegar in the ketchup is not enough.

Bitter
Easy: add ketchup.

So that's the four tastes handled, but there is plenty more to aid flavor enhancement. Ketchup is not all about tomatoes, you know. It contains tomato concentrate, vinegar (sour), high fructose corn syrup (sweet), salt, onion powder, spices (trigeminal), and flavors. The latter may be vaguely described as "natural flavors." Why not use unnatural flavors?

Don't go, there's more. You know presumably that "natural flavors" is a sneaky way of getting you to eat ingredients which may be natural in the same way that arsenic is natural.

Emotion and Memory
The visual image of the fries probably gives us positive, happy memories. If the color produced by the Maillard reaction is what we expect, more the better. If they are gray and limp...not so much.

So far, more than good. What else is there to maximize this fry-eating experience?

Fats
Cooking oil. If you are a cardiologist, leave the room. McDonald's used to use 7 percent cottonseed oil and 93 percent beef tallow. In the 1990s, the anti-science/cholesterol nonsense took hold of facts and the public's attention. Massive financial lobbying by part of the food industry wanted to sell more margarine and seed oils at any cost. Saturated fat was vilified, but that has not stood up to the pressure of facts and science. Vegetable oil pleased farmers because they could get subsidies to grow them and a ready market to buy them. If you want the downside, read Dr. Robert Lustig's *Metabolical*. What does he know? He's just retired professor emeritus of pediatrics in the division of endocrinology and the institute of health studies. He's also not very interested in money.

The funny thing is that beef tallow has a lot of good going for it, and that comes from me as a reluctant vegetarian. What else is now in McDonald's fries? Gums, starches, emulsifiers, and stabilizers. Everything but food, you may snarkily posit. What's the deal with seed oils? Well, they are great if you want type 2 diabetes, stroke, heart disease, cardiovascular disease, cancer, inflammatory disease, and Alzheimer's, all bundled under insulin resistance, metabolic syndrome, and other reasons that will cut your life short in length and quality. Had enough? But beef tallow...mmm. Surely there is nothing else we can add to fries to make them even more irresistible. How about fat in the form of cheese? What the heck, put it on a hamburger. Cheeseburger and fries. The hamburger bun has a soft coziness supporting chewiness.

What have we amassed so far? A half day's potential energy. I don't use the word "calories," because that term is meaningless unless you define which calories and from whence, they came.

Hold on; now you tell me all this talk is making you thirsty. Plain water? Fine, but you can always carbonate it to excite the trigeminal nerve even more with its carbonic acid. Now you are getting sensory overload, aren't you? The stretch receptors in the stomach think you have done well and so signify satiety. "I couldn't eat any more. Satisfied at last." Hold on there; satiety from one set of flavors doesn't ensure satiety to others. There is something called "the Thanksgiving effect," or, "Christmas dinner effect" in the UK. There is always room and desire for a little more, as long as it is different from what you just gobbled down.

"Could I tempt you with a chocolate-peppermint square?" Despite failing in my quest to have been a volunteer in chocolate research, I can reveal the research process. Every five minutes, eager volunteers (minus me) were given a square of chocolate. Yummy, you may say. That is only the case until nausea set in. Some felt repulsed after sixteen squares. Others after seventy-four squares. How short the journey from yummy to yuck can be. Perhaps there is too much of a good thing after all. That's it then finished. No?

How about finishing off with a shot of bitter espresso to contrast with the foregoing? Oh, come on; you must have room for coffee with its six hundred volatile olfactants (smells). If you wish to live a long and healthy life, perhaps this is the time to promise not to laugh at people like me who—still, for the moment you are satisfied, though not necessarily with your fatty liver and fat gut. I had a thought. Why have I never seen an obese centenarian? Obesity is just a symptom, not the problem. Time to read *Metabolical.*

Corked Wine

Nowadays, corks are made of man-made synthetic materials—plastic. Historically, other materials were used, including cloth, leather, clay, glass, and wax. Organic corks come from cork oak trees, which were abundant in Portugal. Like most things organic, they can get infected by organisms. Not just corks, but wine barrels and even toenails. Here, we are concerned with fungi regarding cork. The fungi infect the wine and produce a flavor which is undesirable, wouldn't you think? People perceive it differently, hence "corked". For some reason, wine lovers don't like to spoil their evenings sipping a wine and bringing tetrahydropyridine (TCA, or THP hereafter) into the conversation. This would mean that not just the wine is spoiled, but also the ambiance of the gathering. A wine authority posted this: "The flavor is detected toward the end of the swallow, and the aftertaste can remain for a few minutes. Thirty percent of winemakers are not able to detect the flavor of THP in wine (we do not know if this statistic is reflected in sour beer, but some people have reported not being able to taste THP in sour beer).”

Allow me to dissect the foregoing knowing that you can envisage me laughing to myself throughout.

“The flavor…end of swallow.” Just like everything else we consume, flavor comes after swallowing. Is it possible to detect flavor before the swallow? That would be smell and taste, not flavor.

“Aftertaste can remain…” Just like most things we consume. He’s not exactly breaking new ground, is he?

“Thirty percent of winemakers cannot detect…” In other words, his article will have no relevance to most of humanity and questionably to some select brewers. Who is he writing his article for?

Flavor perception is not chemistry, and understanding chemistry will not take you far in understanding it. TCA also appears in beer as an off flavor.

Gerry’s Kitchen-Science Methodology: The Five Rules

What’s my point?

Humans are absolutely hopeless at correctly identifying items in a mixture.

Charles Darwin apparently spent much fruitful time at his house and garden doing experiments and researching that which was plainly, and accessibly, around him. He did not need a research laboratory at a learning institution. He simply did it at home. Following in his shoes, which are in every respect bigger than mine, I did this utilizing common liquids. Eventually, I branched out to testing friends and strangers at my local pub, My Place. It pleases me that no one mocked me, and they were generally eager to be tested, especially the owner Mark and his sidekick Sharky, who are both recognized beer judges. If I say that I blind-tested everyone, you would be permitted to challenge me with, “So what is so special about that?” Settle down, and I will tell you.

Blind Tasting: Define Blind

In the popular nerdy world of wine, beer, coffee, and other tasting, efforts are made to select the better and best. This is big business and helps finance beer, coffee, wine, and other organizations. Professionals and otherwise like to be judged for reasons of self-esteem and product promotion. Competitions are generally closed-door events, which in the past I used to think prevented ne’er-do-wells from interloping and securing a free drink ride.

So, what is the process? Let us use beer as our example, although broadly, the same applies to wine, coffee, honey, and many other liquids, including olive oil. Judges, usually in small groups of two or more, sit round a table and nervously chew on dry bread to cleanse their palates, which it in fact does not do. Don’t tell them, they are nervous, remember, and it is better than watching them picking at their fingernails.

The steward brings a pitcher of beer and declares its identifying number. The judges have a reference sheet, which expands on the number. For example, #37 fruit ale with raspberries. There are guidelines for that style of beer. The judges pour their samples and study the head and whatever else they deem interesting. But hold on; they are looking! This is supposed to be blind tasting. According to the Oxford English Dictionary, blind means, "Done without being able to see; concealed, closed, or blocked off." I warned you at the beginning of the book that it was in part about words and their meanings, so don't harrumph at me like that just because you may have made a nincompoop of yourself. I wonder who the first nincompoop was?

Gerry's Blind-Tasting Methodology

Rule #1: Don't let the taster know what the liquid is.

It is very runny; that is all they get. It could be any liquid, including water. I often test them with water just to let them know that no, they don't know what is coming. You do not trust me? I understand. Why don't I give them more information on the sample? Settle down and let me tell you a true story. I gave a presentation to some of Connecticut's top wine sommeliers. They identify themselves by arriving with cryptic cloth bags containing secret bottles of wine. Possibly great, but decidedly expensive. Here is how I presented one sample:

"Gentlemen, this next sample is a white wine. Although not infected or spoiled, it is simply not a very good white wine. How would you advise the vintner of this white wine to improve his white wine?" So yes, you got it right—I repeated white wine four times. They followed my instructions and wrote copious notes regarding this white wine. They were all wrong. It was not even wine. No, I am not going to tell you what the sample was short of revealing that there were no grapes therein. However, their brains were only considering white wine without other options.

Rule #2: Do not let them sniff the sample. When it is swallowed, backwash will send smell molecules up to BobHo via the back route. Sniffing just distracts them and wastes time.

Rule #3: Using an opaque syringe (correct, without the needle. Do you really believe that this is the result of an unhappy previous incident?), I draw a 3 ml sample and squirt it directly into the mouth (theirs not mine).

Rule #4: They are instructed that the sample is nothing they haven't experienced many times before, no tricks with strange fruits and spices or small mammals. If they can't identify the sample immediately, then they will be assumed to be guessing. Just swallow and identify, folks. Don't overthink it.

Rule #5: Sometimes with a small group (under twenty-five), in order to intimidate and shock them, I'll offer a sample with the proviso that if they can identify it, then they must put ten dollars cash on the table and I will do the same, betting on their failure. That tends to cut things short. No one has ever taken up my challenge, which tells them that they don't know what they are talking about from the start. Having said that, one taster, Clay Viands, did not bet but got the two-item sample correct. That made him one out of twenty-five. No ten dollars for him though. Let me say here that he is very, very, very unusual, from my experience. If there was an Olympic contest for tasting, he'd probably get the gold.

Get to the Point, Gerry; What's Your Takeaway?

Hold your horses. I will not analyze all my results, because they are not based on a double-blind trial. They are based on common sense in real-world situations with a cross section of tasters. You are invited to sneak into my tasting process at a few defined events, some small. For example:

A short, balding man walks into a bar and says, "Would you open your mouth and I'll squirt a liquid in and then you tell me what it is?" Sharky the barman interjects, "It's okay, you are safe—it's just Gerry." MY PLACE management was kind enough to let me take their back dining room to lay out twenty-one test samples. They were liquids, each with only a number identifier. Each taster had a score sheet where they logged their opinions. In an effort to get to the point, I will reveal that very few got anything correct. It was interesting what they got wrong. The first sample was a mix of 60 percent Cabernet Sauvignon and 10 percent each of lemon, orange, vinegar, and vanilla. Vanilla was the only one that was correctly identified and even then only a few times. An amazing variety of suggestions was given; I will only show you those that were wrong/invented.

Gerry's Taste Panel Composite Results from England (Castle Rock Brewery) and Newtown, Connecticut MY PLACE)

1. 60 percent Cabernet Sauvignon + 10 percent each lemon, orange, vinegar, vanilla = chocolate, vanilla, wine, raspberry, Creamsicle, cocoa, medicinal, red wine, roasted grains, black cherry, mulled wine, blackcurrant, horseradish, dishwashing liquid, salt, fish (!), chili

2. Lime = lemon, lime, ginger, sour mix, lavender, key lime, grapefruit

3. Lime + lemon = lime, vanilla, vinegar, lemon, burning

4. Lemon = lemon, vanilla, more acid than #3, lemon thyme, grapefruit

5. Key lime = lime, lemon, vanilla, lemon concentrate, astringent, key lime, pineapple

6. Meyer lemon = pineapple, vinegar, lemon, lime rind, sweeter lemon, lemon custard, grapefruit, orange, vegetal

7. Meyer lemon + lemon zest = pineapple, tropical fruit, mango, orange, lemon, lime, lemongrass, tangerine, grapefruit, lovely

8. Blood orange = raspberry, strawberry, vanilla, pineapple, apple, prune juice, sweet pepper, tomato, orange

9. Orange = orange, papaya, pineapple

10. Orange + blood orange = apple, orange, raspberry, strawberry, grapefruit, vegetal

11. Mandarin + tangerine = (tainted sample—too dilute)

12. Blood orange + orange zest = strawberry, raspberry, orange, all-spice, watermelon, carrots, vegetable, all-spice, ginger, horseradish, lovely, intense citric, passionfruit

13. Strawberry + lemon + lime (as in, strawberry daiquiri mix) = lemon, lime, strawberry, lime rind, watermelon, thyme, blood orange, vinegar, mango, nectarine

14. Blackcurrant = blackberry, tart berry, mango, beets, raspberry, kiwi, guava, plum

15. Blackcurrant + blackberry = blueberry, blackberry, orange, pineapple, raspberry, vegetables, vanilla, not much, pomegranate

16. Blackberry = lemon, lime, blueberry, orange, beet juice, tart dark fruit, raspberry, not much, graham cracker

17. Lemon + orange + vinegar = pineapple, lime, lemon, vinegar, tart vanilla, Hawaiian punch, acetic acid, ginger, nasty

18. Astringency test with 1 percent alum = astringent, salty, nothing, alum, chalked, minerals

19. Metallic test with 1 percent potassium chloride = vegetal, bitter, nothing, smooth, full, metallic, salty

20. (Strawberry daiquiri) Rum + lime + strawberry + sugar = alcohol, strawberry, lemon, too much alcohol

21. (Strawberry daiquiri) Rum + sugar + Perfect purée + strawberry (too much rum) = big strawberry, too much alcohol, better than #20

The takeaway is:
- Zest seems to potentiate the flavor.
- Vinegar is not easy to detect.
- Strawberry daiquiri is better with less alcohol and no lemon/lime.
- Metallic remains the mystery you are tired of me talking about.

BevCon 2017 Flavor Presentation Test

The mix I used was the same as for the test above—wine, lemon, orange, vanilla, and vinegar. Of most interest was the number of items identified, which were pure imagination, and these were from flavor professionals, remember.

Lime	Cinnamon
Tangerine	Baby aspirin
Grapefruit	Passionfruit
Rotting grapefruit	Peach
Tang	Coffee
Guava	Bloody Mary
Hi-C	Cherry pie
Champagne	Strawberry
Playdough	Ginger
Melon	Tomato
Creamsicle	Banana
Caramel	

Aspetuk Brew Labs Off-Flavor Workshop

Forgive me for taking advantage of the event. I snuck in some of my own samples but was tasked to explain the biological implications of the most common beer off-flavors. There were twenty-five attendees, all experienced tasters.

Lemon and orange mix—zero correct.

Lemon and apple cider vinegar—zero correct, except Clay Viands, who identified both correctly.

The foregoing is to give you a taste of how hopeless humans are at correctly identifying items in a mixture and at the same time identifying items that are not there. In the fermenting industry, especially wine and beer, vinegar (ascetic acid) is a spoilage indicator, but even at relatively high concentrations, it is hard to detect. Lemon is much easier, but vanilla takes the cake.

DEEP THOUGHT: Can flavor exist without memory and emotion?

Food for the Intellectually Curious

- Penn & Teller (everything they do)

- *Metabolical* by Robert Lustig, MD, a doctor disillusioned with our medical system which does not address health—only money-making disease. Reveals flimflam in modern medicine.

- *Outlive* by Peter Attia, MD. Science and art of longevity. See under Robert Lustig.

- *The Food Lab* by J. Kenji López-Alt. Science versus flimflam in cooking.

- *Breath* and *Deep* by James Nestor. Original thinker. Makes sense of why and how we breathe and a generally interesting read. He got me mouth-taping at night.

- The Skeptics Society and *Skeptic* magazine. Rife with magicians and questioning everything.

- Pradip Jamnadas cardiologist, educator, and freethinker. Ah yes, he is funny.

- *Annoying: The Science of What Bugs Us* by NPR Science correspondents Joe Palca and Flora Lichtman. Find out how annoying you are; take the free assessment by the authors and Drs. Robert Hogan and King of Floonia: https://youbugme.perfprog.com/

Does Smell Exist? Nerd Alert

If you think I am crazy, then you have not kept up with the latest in physics. Does time exist? Do we exist? What is consciousness?

Seventy-eight percent of air is nitrogen. Does nitrogen smell? Is it odorless? No? How do you know? What if 50 percent of air were garlic smell? Would you accommodate to it in the way that if you have a garlic-heavy evening meal and wake in the night. Can you smell it in the same way that your bed partner says she/he can? Garlic smell lingers for as long as you are not swamped with it. Perhaps nitrogen has a smell, but because we are flooded with it twenty-four hours a day from the moment that we are ejected from the birth canal. I don't personally know the answer so decided to get the opinion of an astrophysicist who might shed some insight. Full disclosure: just like dark matter, he asks just to be identified as a shy astrophysicist...

Hi Gerry,

That's an interesting theory! We do become desensitized to smells that we are exposed to for long periods.

While nitrogen has no smell that we can detect, several nitrogen compounds have noxious odors. It's probably because some nitrogen compounds are harmful, so there's an evolutionary benefit in smelling them.

A double-blind study would be possible using atmospheric chambers, with pure oxygen in one and nitrogen-rich air in the other. You could even make a breathable atmosphere of oxygen and helium, but that would make it obvious to the test subject as to which chamber they were in.

The Navy used oxygen and helium in their undersea laboratories in the 1960s. I wonder what fresh air smelled like to the aquanauts after breathing that for months.

My guess is that humans have no smell receptors to detect pure nitrogen because there is no evolutionary benefit. If any of them are still alive, I'd ask an early aquanaut.

Notes

Record here the items with which you disagree. Now you have enough for your own book, eh? Gerry is certainly very good at repeating himself.

Ultimately,

everything comes down to evolution.

"You say that you don't like how hominins are evolving; so don't eat them!"